D1583089

rice trails

*Rice Trails – A Journey through the
Ricelands of Asia & Australia*
February 2004

Design: Bill Burrows
Digital Imaging: Ryan Evans
Typography: Rob Walker
Mapping: Tony Fankhauser, Wayne Murphy & Paul Piaia
Editing: Lindsay Brown & Jenny Brown

Published by:
Lonely Planet Publications Pty Ltd ABN 36 005 607 983
90 Maribyrnong St, Footscray, Victoria 3011, Australia
www.lonelyplanet.com
AOL keyword: lp

Printed by the Bookmaker International Ltd
Printed in China

Photographs:
The images in this book are available for licensing from
Lonely Planet Images.
www.lonelyplanetimages.com

ISBN 1 74104 309 3
Text © Lonely Planet Publications Pty Ltd 2004
Photographs © Richard I'Anson 2004

Lonely Planet Offices:

Australia
90 Maribyrnong St
Footscray
Victoria 3011
tel: 03 8379 8000
fax: 03 8379 8111
email: talk2us@lonelyplanet.com.au

USA
150 Linden St
Oakland
CA 94607
tel: 510.893 8555 toll free: 800 275 8555
fax: 510.893 8563
Email: info@lonelyplanet.com

UK
72-82 Rosebery Ave
London EC1R 4RW
tel: 020 7841 9000
fax: 020 7841 9001
email: go@lonelyplanet.co.uk

France
1 rue du Dahomey
75011 Paris
tel: 01 55 25 33 00
fax: 01 55 25 33 01
email: bip@lonelyplanet.fr

rice trails

A Journey through the Ricelands of Asia & Australia

Words Tony Wheeler

Photographs Richard l'Anson

Clockwise from top left:
Philippines – *A smiling villager at Batad, one of the most picturesque rice growing centres in the highlands of Luzon.*
Vietnam – *Bay Tinh sells rice at the floating market in the Mekong Delta town of Cai Rang.*
Philippines – *It takes an hour or two to walk from the main road across the hills and down to Batad with its famous rice terraces, we met Tagtagon Bumanghat beside the trail.*
Bangladesh – *A worker at a rice processing plant in Latifpur, north-west of Dhaka.*
Bali – *At a temple ceremony at Pejeng, just outside Ubud, a priest has rice grains on his forehead.*
Burma – *A farmer in the village of Tha Yet Khon.*

Clockwise from top left:

China – A rice farmer at the picturesque Dragon's Backbone Rice Terraces near Longsheng.
Nepal – Usha is a young Nepalese girl, working with her mother in a rice field just outside the Kathmandu Valley
Bangladesh – A woman in the rice market in Dhaka.
India – Sifting the rice at the Khanna grain market in the wealthy Punjab state.
Bangladesh – A pause for a cigarette between unloading sacks of rice at the riverside in Dhaka.
China – Her traditional outfit reflects the traditional nature of rice farming at the Dragon's Backbone Rice Terraces.

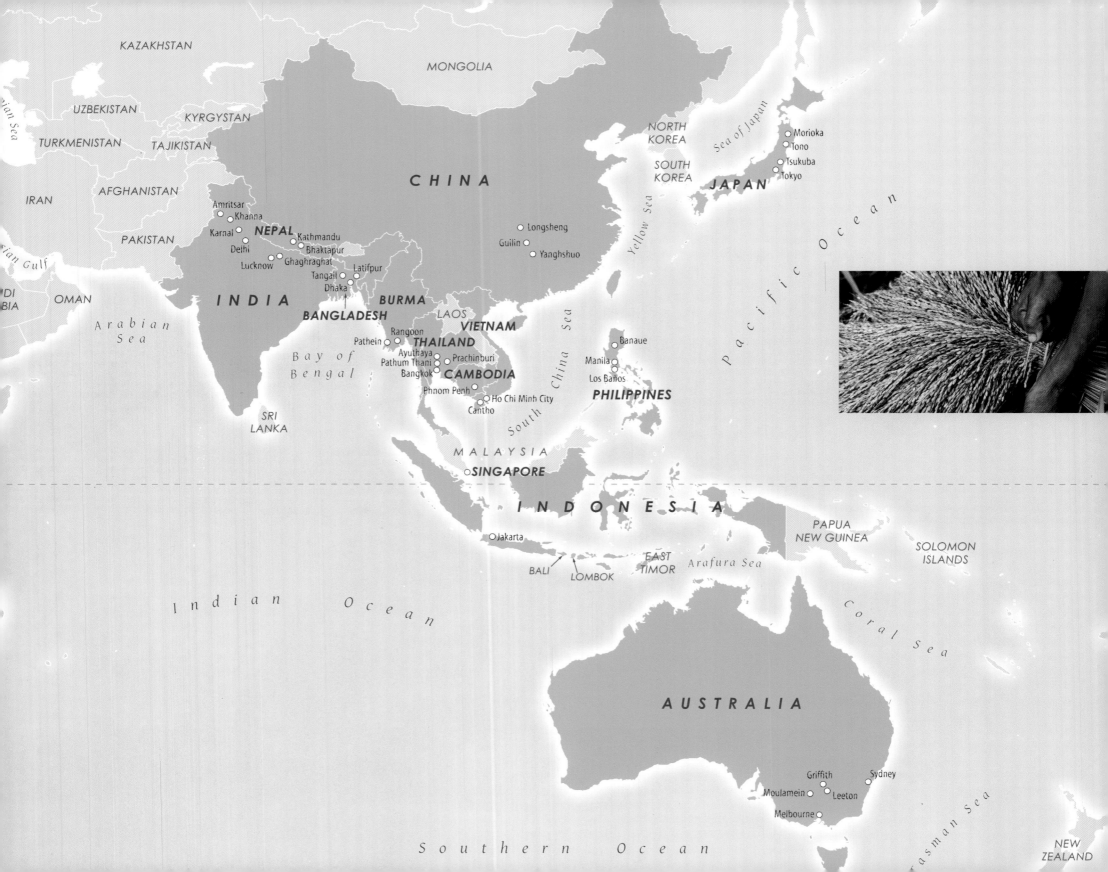

KAZAKHSTAN

MONGOLIA

UZBEKISTAN

KYRGYSTAN

TURKMENISTAN

TAJIKISTAN

NORTH
KOREA

Sea of Japan

Morioka

CHINA

Tono

Tsukuba

SOUTH
KOREA

JAPAN

Tokyo

IRAN

AFGHANISTAN

PAKISTAN

Amritsar

Khanna

Karnal

NEPAL

Kathmandu

Longsheng

Guilin

Yanghshuo

Delhi

Bhaktapur

Lucknow

Ghaghraghat

Latifpur

Tangail

Dhaka

Yellow Sea

Pacific Ocean

DI
BIA

OMAN

*Arabian
Sea*

INDIA

BANGLADESH

BURMA

LAOS

VIETNAM

*Bay of
Bengal*

Pathein

Rangoon

THAILAND

South China Sea

Banaue

Ayuthaya

Prachinburi

Manila

Pathum Thani

Bangkok

CAMBODIA

Los Baños

Gian Gulf

SRI
LANKA

Phnom Penh

Ho Chi Minh City

Cantho

PHILIPPINES

MALAYSIA

SINGAPORE

Indian   Ocean

INDONESIA

PAPUA
NEW GUINEA

SOLOMON
ISLANDS

Jakarta

BALI

LOMBOK

EAST
TIMOR

Arafura Sea

Coral Sea

AUSTRALIA

Griffith

Sydney

Moulamein

Leeton

Melbourne

Tasman Sea

Southern   Ocean

NEW
ZEALAND

rice trails

Introduction

I can pinpoint exactly where and when I first fell in love with rice. It was in East Java in '74. I was in my mid-20s, recently married, and my wife Maureen and I had been travelling through Indonesia for several months. I'd certainly seen (and eaten!) lots of rice by that time. I was familiar with beautiful rice terraces in Nepal and many other Asian countries, but in the country around Yogyakarta everything came together. Perhaps I'd had sufficient grounding to really appreciate the beauty of rice; perhaps it was simply the right season, the right light, the right blue sky to contrast with the green rice fields.

Whatever the cause the Indonesian rice fields were uniformly sublime: incredibly lush and green, or mellow and gold, or every shade in between. The way the rice stalks bent, that graceful bow as if they were politely inclining towards you, bowing deeper as the rice grains grew heavier, closer to harvest. In every batik gallery in Yogyakarta the images reflected the same Arcadian simplicity – rice in its multi-green-coloured glory, the rice stalks bent serenely under their burdens, the rice farmers standing on the paddy bunds, regally surveying their kingdom.

Thirty years have passed since that rice paddy epiphany and a great deal of rice has also passed by. With friends I've followed the pilgrim trail across the Himalaya in western Nepal to Tibet's Mt Kailash, accompanied all the way by goat trains packing small leather bags of rice to be traded for salt. With my early rising baby daughter in my arms I've watched the sun greet the picture-perfect rice paddies in Ubud, Bali. Countless *thalis*, the classic 'rice plate' vegetarian meal of south India, have sustained me on journeys through the subcontinent. I've breakfasted on rice in Japanese *ryokans* and savoured it in various national dishes from *chelo kabab* in Iran to *khao phat* in Thailand. I've eaten it with forks in the west, spoons in the east, chopsticks in China and fingers in India.

Rice is undoubtedly humanity's most important food (everyday more calories of rice are consumed than anything else) but it's also the most beautiful crop on earth. Whether it's rice paddies tripping down a hillside in the Philippines, China or Indonesia, each step a little masterpiece of artistic perfection, or sweeping across laser-levelled bays in pancake-flat Australia, it's always beautiful. Nor does the story end with taste and visual appeal; every step of the way from planting, through growing, harvesting, processing, trading and consuming has its own interest and colour.

Furthermore, the research which led to today's high-yielding 'dwarf' varieties was a keystone in the 1960s 'green revolution', the great leap forward which banished the centuries old spectre of famine in Asia.

The English word for rice comes from the Italian *riso* which in turn derives from the Latin *risium* and the Greek *oryza*. Science classifies rice as belonging to the genus *Oryza*, either *Oryza sativa* (the familiar rice of Asia) or *Oryza glaberrima* (the less common red rice of West Africa). Rice in its rough, unprocessed state, prior to the milling which removes its tough outer covering, is known as *paddy*. The word is derived from the Indonesian/Malay word *padi* and has been adopted into English as the word for a rice field, a rice paddy.

Of course a food as important and ubiquitous as rice has earned its own lexicon. Just as Eskimos have 100 words for snow and Tibetans dozens of words for yaks, the Indonesians have many precise descriptions for our one word rice. To these rice specialists it's only *padi* while it's still in the field.

Once it has been harvested, but not yet husked and milled, the rice becomes *gabah*. Once it's milled, but still not cooked, the rice is *beras*. Rice in its cooked form is *nasi*, hence that most popular Indonesian dish *nasi goreng*, fried rice. A rice field, a paddy field in English, is *sawah*, or at least it is in its familiar flooded state, a dry rice field is *ladang*.

The Japanese are equally specific with their definitions and can conjure up at least 30 different words for rice. Rice itself is *gohan* in Japanese; breakfast is simply morning rice, *asa gohan*. Similarly in China the traditional greeting is simply, 'have you eaten rice today?' And a Mao-era job-for-life was the 'iron rice bowl.' In rice-loving Thailand a farmer is a *chao naa*, literally a 'rice field person'. Rice is so central to the countries of Asia that even familiar words like Honda and Toyota have their origins in rice terminology.

Rice may grow all over the world but 90% of it is grown and eaten in Asia. It is in Asia where rice was first cultivated and where rice today is overwhelmingly the number one food. With a small excursion to nearby Australia, principally to look at rice growing in a very different context, the rice trail celebrates rice in its homeland, Asia.

Indonesia – Near the village of Batubulan in Bali, not far from Ubud, the rice shoots have recently been transplanted into a flooded field.

History

Today rice is grown on every continent except Antarctica, it is a staple food in many countries, a key element of many national cuisines (rice in risotto is just as Italian as pasta in spaghetti) and a major export crop in countries like Australia or the USA, countries where rice is definitely not an all important part of the diet. Nevertheless it is Asia where rice has its real home. Rice is the principle food for the world's two biggest countries – China and India – which between them account for more than half the world's rice production. Add Indonesia and Bangladesh to the picture and you've accounted for two-thirds of the world's rice; throw in the rest of Asia and the figure rises to 90%.

The closest rival to rice as a food product is wheat. In fact there is probably slightly more wheat grown each year than rice's 500-odd million tonnes, but a significant portion of the wheat crop, perhaps 20%, is fed to animals. Corn, another 500 million-tonne-a-year crop, is used even more heavily as animal feed. In contrast, rice is almost all eaten by humans.

Rice comes in two main forms, *Oryza sativa* and *Oryza glaberrima*, but *O. sativa* is by far the more common, accounting for well over 90% of the rice grown worldwide. There are numerous other much less common species of rice, but wild rice, popular though it is as a gourmet speciality, is not really a rice at all. Attempts are being made to cross cultivated rice with wild *Oryza* species to develop a new variety of cultivated rice with increased resistance to diseases and pests.

Within those two main types of rice are a huge number of subdivisions from the fragrant, long-grained and highly esteemed Jasmine rice grown in Thailand to the slightly sticky rice favoured by the Japanese. Of course every nationality has its own favourite rice, but objective rice connoisseurs often give the accolade to the Basmati rice of the Punjab region shared between India and Pakistan.

Philippines – At the Pahiyas Festival in the village of Lucban, south of Manila, stalks of unthreshed rice, rice grains, circles of bright red kiping (edible decorations made from brightly coloured rice flour wafers) and chilli peppers are decoratively displayed.

Humankind's transition from hunter-gatherers to farmers commenced about 10,000 years ago. It didn't happen in just one region; farmers appeared on the scene in various places around the world, but what they grew varied from one place to another. In Europe wheat was the key early crop, but in Asia it was rice and the greater productivity of those early rice farmers was a prime reason for the rapid development of early Asian civilisations. The population of China during the Confucian era around 500 BC was 10 times that of the Athenian Greek civilisation.

Clearly rice growing dates back to the very earliest development of agriculture, and today it is generally agreed that the techniques of growing rice in flooded fields originated in China and later spread to other countries in Asia. In the Yangtze Basin in China there is evidence of rice consumption during the Neolithic period from 6000 to 9000 BC. Rice grain imprints have been found in pottery shards dating from 4000 BC in Thailand. Archaeologists have determined that rice was an important food for the Indus Valley civilisation in Moenjodaro, in modern Pakistan, by 2500 BC. Rice was taken back to Greece and other Mediterranean countries by Alexander the Great's forces around 344 to 324 BC.

High yields ensured the early success of rice and its extraordinary adaptability led to a wide distribution and subsequent divergence. The warm, wet tropical regions of Asia may be rice's true homeland but rice grows beyond 50°N in Russia and to 40°S in Argentina. It is found high in the Himalaya in Nepal and Bhutan, yet surfers make their way through the rice paddies on the way to the beach at Kuta in Bali. The extraordinary deepwater rice of Bangladesh can flourish in water up to five metres deep but rice also grows in the dry deserts of Iran, and the flat plains of Australia's semi-arid western region of New South Wales produce the world's highest rice yields.

Clockwise from left:

Philippines – *Carved images of bululs, the primitive looking rice god figures which are set up to guard rice granaries in the Ifugao ricelands of North Luzon.*

India – *Offerings laid out for a wedding ceremony inevitably include rice.*

Indonesia – *At Ceking, to the north of Ubud in Bali, offerings are scattered by the rice paddies. These decorative gifts are intended for demons, bribes to persuade them not to make their way into the rice fields.*

Ricescapes

It is often said that rice is the world's most beautiful crop; and admiring the stunning rice terraces of Bali in Indonesia or Banaue in the Philippines it's easy to appreciate where that view originated. In fact rice is grown in four main eco-systems, and those hillside rice terraces are simply part of the most common system.

Rice terraces are the best known example of irrigated rice farmland, a group which also includes the flat, high tech, rice fields of the USA and Australia. Irrigated fields not only account for the largest percentage of riceland, about 65% of the total world area devoted to rice, but they are also the most productive. Despite the already high yields, it is from irrigated rice fields that scientists see the most potential for future improvements. It comes down to control; with precise control over the amount of supplied water the farmers can provide the rice with optimum growing conditions.

The second most important rice ecosystem includes the rain-fed lowlands which account for about 25% of total rice production. Here the crop depends upon rain arriving to schedule, and in sufficient – but not over-generous – quantities. The monsoon regions of Asia, where the rain can usually be depended upon to arrive with clockwork regularity, are the true home of this ecosystem. The problem with rain-fed rice is its total dependency upon nature playing its part. If the rain is late or inadequate, the seedlings may die before they've established themselves. Too much can be just as disastrous as too little; torrential rain may simply wash the seedlings away.

The remaining 10% of the world's rice production is split between two less productive ecosystems. In some lowland regions of South and South-East Asia, particularly in Bangladesh but also in India, Burma and Thailand, the monsoon rains come with such intensity and the land is so low-lying that some areas may be deep under water for months at a time. Amazingly, there are varieties of rice that can cope with these conditions, growing up to 25cm (nearly one foot) a day as the flood waters arrive and thriving in water up to five metres deep! In these conditions the rice has to be tended from boats, yields are not especially high and complete crop failure is not unusual, but nothing else will grow there.

The final rice-growing areas are the dry uplands, typically found in the regions populated by the hilltribes of northern Thailand, Laos and Burma. Again yields are very low in these less-than-ideal conditions, and primitive and destructive slash-and-burn farming techniques are often employed. For these subsistence farmers there's little choice, it's rice or nothing.

China – *Near the provincial town of Longsheng in Guangxi province the Dragon's Backbone Rice terraces are some of the most beautiful in the country. Following Pages: *Philippines* – From the regional centre of Banaue in northern Luzon a road winds west through the mountains with the Ifugao province's famous rice terraces always in view. Just south of the road the village of Bangaan nestles in an extensively terraced valley.*

NEPAL

Rice in the Himalaya

Rice will grow up to 3000 metres altitude and in Nepal it's quite common to find rice paddies with snow-capped Himalayan peaks as a backdrop. The monsoon ends in late-September to early-October and as the skies clear and the trails dry out, trekkers and mountaineers head out of the valley and rice farmers start to take in their harvest.

Near the medieval looking town of Bhaktapur, at the eastern end of the Kathmandu Valley, a typical brick home belonging to a rice-farming family overlooks the terraces. The monsoon has run late this year so this is one of the first dry, clear days although it is already well into October, and the farmers and their families have quickly moved out into the fields to bring in their crop.

In the background the mountains of the Langtang Range provide a classic Himalayan backdrop. Sharp-edged Can Chempo (6387 metres) points skyward directly above the farmhouse while Dome Blanc (6830 metres) and Dorje Lakpa (6988 metres) dominate the right side of the view. In fact some of the more distant mountains in view are even higher. Unhappily crystal clear days like this are becoming a rarity as traffic and industrialisation cast a pall of smog across the Kathmandu Valley.

A patchwork of rice terraces in the Pokhara Valley.

Facing Page: A farmhouse and rice terraces are backdropped by the Himalayan mountains.
*Following Pages: **China** – From the major city of Guilin a mountain road climbs up to Longsheng near the famous Dragon's Backbone Rice Terraces.*
A visual feast of rice fields border the road for most of way, here the Zhaung farming village of Da Liu lies close to an endless staircase of rice terraces.

RICESCAPES 21

INDONESIA

Planting & Harvesting at Ceking

Just north of Ubud, the cultural centre of Bali, Ceking is one of the island's postcard-perfect stretches of terraced ricescape. Where two valleys converge the terraces step down the hill in an intricate three-dimensional patchwork, a jigsaw puzzle of interlocking rice fields. Some of the fields are no bigger than a double bed, others as large as a basketball court, and they come in all sorts of shapes: squares, rectangles, parallelograms, with edges straight, angled and gently curving. Coconut palms and banana trees add to the composition.

There are no seasons for rice growing in Bali, every stage from preparing the fields and planting through to harvesting and beyond will be going on at the same time although rarely at the same place. Ceking is clearly an exception and at one glance we can take in the whole process, Bali style, and toothy Nyoman Seprag clambers up and down the terraces with us, pointing out what's going on.

Farmers plough the flooded fields repeatedly, raking them over and smoothing them out to give the earth a uniform muddy consistency. Meanwhile, the rice seedlings are growing in a small nursery field; massed together there's simply no space for weeds to get a foot in the door.

At Ceking, north of Ubud in Bali, postcard perfect rice terraces step down the hill but the earth bunds between the fields require constant maintenance.

The next stage is the methodical transplanting of the seedlings, now about 15cm high, into the paddy fields. Transplanting seedlings at this size gives them a healthy head start over any intruding weeds and means less herbicide will be used. The price paid is the intensive, and back breaking, labour involved in transplanting.

Now it is easy street for the rice farmer, who can sit back and watch the rice grow. Sure you have to keep greedy birds at bay and scarecrows loom over many rice fields, but making offerings in field-side shrines to Dewi Sri, the goddess of rice and harvest, probably helps as well. It takes about three months for the modern dwarf varieties of rice to reach maturity, a much shorter time span than the traditional padi bali, which is still found in a small percentage of the island's rice fields. As the rice grows it changes colour from the seedling's vividly fresh green through a darker green and finally to a beautiful green and gold mosaic. At the same time the growing rice grains, each cluster known as a panicle, weighs down the rice stalk, the tip bending over in that gentle curve that is as much an element of a rice field's seductive attraction as the colours.

At harvest time the villagers move into the fields en masse, for the first time since the seedlings were transplanted months earlier. Using sharp-edged sickles the rice stalks are chopped down and threshed in the fields to remove the rice grains, which are bagged and sent off to be milled. The field, already drying out before the harvest, will now be left to dry completely, then burnt off, flooded and ploughed under before the whole process starts again.

The terrace steps at Ceking vary from delicate little edges just a few cm high to mighty stairs, rising three or four metres from one flooded level to the next. Shrines to Dewi Sri dot the fields; sometimes they're simple affairs about the size of a birdhouse, topped by a tin roof and standing on bamboo legs, other shrines are sturdier creations of stone. Life in Bali is always a question of balance, so bad spirits must be bribed, even as the good ones are honoured. More offerings are scattered derisively along the field edges, payoffs which should keep stray demons from wandering into the paddies.

The fields are full of more mundane life than good and evil spirits. Dragonflies flit over the rice fields or between the maturing stalks. At night children come out into the flooded paddies to look for the tiny eels which are a Balinese delicacy. Tadpoles wiggle around the flooded fields and frogs frogkick across from one side to the other. Lizards, birds and grasshoppers all make an appearance in the rice paddies. The flooded fields, waiting to be ploughed or planted, are also the daily home for flocks of ducks. Led out to the fields each morning and back each evening by a duck shepherd, they're a much loved part of the Balinese rural scenery. All this animal life features regularly in Balinese paintings, wood carvings or stone sculptures, although the real rice paddy frogs are rarely encountered wearing crowns and toting umbrellas.

Top:
Hardworking rice farmer Wayan Kantun.
Middle:
Flowers and insects, there's always lot of life around the rice paddies.
Bottom:
Decorations in a temple to Dewi Sri, the goddess of rice.

PHILIPPINES

Banaue Rice Terraces

In the north of the main island of Luzon, 200km north of Manila as the crow flies but much further on the winding mountain roads, the town of Banaue is in the Ifugao heartland. On the surrounding hills these hardy mountain dwellers have spent the last 2000 years building stone-walled rice terraces which, if stretched out in one long line, would run for 20,000km. It's hardly surprising that they've been dubbed the eighth wonder of the world.

The town itself stands at 1200 metres and has that unmistakable feeling of an Asian hill station. Any visitor familiar with Shimla or Darjeeling in India, Nuwara Eliya in Sri Lanka or Dalat in Vietnam would feel an instant wave of nostalgia when confronted with the town's streets, lined with shabby houses with rust red or faded green tin roofs, stacked down the hillside or balanced on razor ridges.

In 1995, the Ifugao rice terraces were included in UNESCO's (United Nations Educational Scientific & Cultural Organization) World Heritage List as a protected cultural landscape.

The Ifugao rice terraces of North Luzon are on UNESCO's World Heritage List.

Batad Rice Terraces

The village of Batad, surrounded by its neatly terraced fields, is clearly a contender for the title of world's most beautiful rice terraces. It's a natural amphitheatre with rice terraces for the rows of seats and the village itself as the centre stage. Many of the buildings are still thatch roofed while the church and the other, generally newer, buildings sport red tin roofs, which contrast with the glowing green of the rice paddies. Tall betel palms dotted amongst the buildings add the final exotic touch.

Most of the small inns which cater to visitors are high up on the edge of the amphitheatre, overlooking the wide sweep of rice paddy 'seats' tripping down to the stage. Their simple little ply-walled rooms (there's no electricity and fairly rudimentary plumbing) cost just a few dollars a night. Sitting on the terrace of one of the inns, sipping an afternoon cup of tea as a fierce rainstorm sweeps across the varied shades of green, it looks (and it is) positively idyllic; the stone-faced terraces neatly contour the hillside, the rice field green is perfect, all's well with the world. The lucky farmers of Batad enjoy a dual income; rice production is backed up by a steady flow of tourists, flocking there to admire these beautiful terraces.

But will it stay that way? When the terrace walls collapse, as inevitably they do from time to time, repair and reconstruction is no longer immediate. Recently it's only been government aid which has finally prompted repairs to some of the damage. Will the magnificent theatre be dismantled seat by seat, row by row? Will the young people of Batad abandon the rice fields for jeepney driving and bar dancing? Will the steady flow of tourists, walking for two hours over the hills from the main road, only be an interlude, a mere delay in the inevitable decline? It's hard to be hopeful, for pretty though it is, rice growing in Batad is neither efficient nor cost-effective, and if the rice fields go, the other half of the village economy – rice field tourism – will quickly follow.

The village of Batad is surrounded by stone-walled rice terraces.

AUSTRALIA

Paddies to Bays

Although in scale and terrain Australian rice fields are very different from their Asian counterparts they can still be a treat for the eye. Where Asia has rice paddies Australia has rice bays, either contour bays or laser-levelled bays. Contour bays are built along natural contours of the land, segmenting off areas of land of equal height above sea level. Because they are not perfectly levelled they are not as water efficient as laser-levelled bays and they do not allow the water to flow evenly. Today contour bays are not used much, only by farmers who haven't invested in laser levelling technology. Laser levelling costs about US$250 a hectare but doesn't have to be done again for 10 years.

From the air the contour bays of rice fields form graceful curves near Leeton in New South Wales.

Recently harvested laser-levelled bays near Leeton.

BANGLADESH

Deepwater Rice

Fly over Bangladesh towards the end of the monsoon season and it's very clear just how much of the country is underwater. So it's hardly surprising that this is the last stronghold of deepwater rice. Even in Bangladesh this traditional variety is on the retreat, but with so much of the country submerged from June through October there's clearly potential for rice that can cope with not just the few cm of water which most rice varieties quite like, but water up to five metres deep!

Deepwater rice is typically sown in April and needs at least 15 days growth to establish itself before the floodwaters arrive. Once it has got going deepwater rice can typically grow up to 10cm a day; some varieties can even manage 25cm daily growth in order to keep up with the deepening floodwater. By August the rice is simply floating, until the water recedes with the end of the monsoon in October. The rice will be harvested in November or even December.

Yields of deepwater rice are low, usually around a tonne per hectare, but once the monsoon rain appears nothing else is going to grow. Nor is there a lot of work involved with deepwater rice; the seed are broadcast rather than transplanted and once it is growing there's not much weeding, fertilising or thinning and definitely no irrigating to worry about. Furthermore, insist the deepwater enthusiasts, the rice tastes better – sweeter – than any of the modern varieties.

Farmers punt out to inspect their crop, passing through the same narrow channels that locals use to ferry across to their homes, isolated on little islands for up to six months during the wet season. Traffic – buses, autorickshaws and the ubiquitous rickshaws – passes by on causeways and bunds and there is a lot of waterside activity. Big counterbalanced fishing nets are lowered from the roadside into the water and waterlilies and water hyacinth decorate the edges.

A deepwater rice farmer punts out to inspect his crop, near the village of Mohammedpur not far from Sonargaon to the east of Dhaka.

Indonesia – Hotels along the Monkey Forest Road in Bali's cultural capital of Ubud advertise their rice paddy outlooks but in recent years low key guest houses have given way to fancier hotels and swimming pools while the rice paddies have been in steady retreat.

Rice Tourism

Rolling fields of wheat in the American mid-west may be picturesque but nobody has ever built a hotel where views of the wheat fields are the main attraction and you've been drinking too much Guinness when you start dreaming of stunning views of a field of potatoes in Ireland. In Asia, however, the rice paddies really can be a tourist attraction. In Bali the astonishingly beautiful (and also astonishingly expensive) *Amandari* incorporates the rice paddies tripping down from the hotel to the Ayung River below as a key part of the ambience. The nearby, and equally expensive, *Four Seasons Sayan* incorporates some artistically sited rice paddies into the hotel's garden design. The hotels along the Monkey Forest Road in the nearby village of Ubud all look out over rice paddies and even claim 'rice paddy views' in their advertising.

It's a similar story with the beautiful Long Ji or Dragon's Backbone Rice Terraces in the Chinese province of Guangxi. A few years ago the small village of Ping An was just a simple farming village but already a sprinkling of rock bottom guest houses have opened to cater for western backpackers. The first larger hotel soon followed and, unhappily, a horribly destructive road has been chopped into the hillsides to bring busloads of Taiwanese day trippers up to admire the view.

China – In the village of Ping An a new hotel overlooks the famed Dragon's Backbone Rice Terraces.

Growing

Whether the rice is being grown in a tiny hillside terrace in Bali or in a field running flat to the horizon in Australia, the growing cycle, like any crop, is essentially the same.

Water is the unchanging element in growing rice. Any crop needs water to grow, but rice is submerged in it. Ensuring the correct volume is at hand is part of the art of rice growing – too little and the rice will die, too much and it will drown or even be washed away. Water is almost always a precious commodity and most rice growers strive to make their water resources go as far as possible. That can mean skilfully trickling the water down a hillside from one terraced paddy field to another, or it can mean flooding irrigation water into a field which has been levelled with laser-controlled precision to ensure the crop is submerged to exactly the depth required.

The first step is the preparation of the field. In many developing countries that can mean repeated ploughing with an ox-pulled plough or the same process with one of the Chinese-manufactured, hand-guided mini-tractors. In advanced countries – or even the more advanced regions of the third world such as in Thailand or the Indian state of Punjab – tractors come into their own. They range from the high tech, but tiny, devices unleashed on equally tiny rice fields in Japan to the large tractors used in Australia. The ploughed field is flooded and ploughed again to turn it into a smooth slurry prior to planting.

At this stage two very different procedures are followed. In much of developing Asia, rice seedlings are carefully nurtured in a miniature nursery field, then dug up, divided out and transplanted in neat rows. There are a number of reasons for this laborious practice; one of the most important is that the transplanted seedlings have a head start on any competing weeds. With luck they'll continue to outgrow the competition and there will be much less need for weeding and the use of herbicides. Even in advanced rice growing countries like Japan, farmers may follow the transplanting process, although sophisticated machinery does the backbreaking work of planting out the individual seedlings.

In other environments the rice may be directly seeded or 'broadcast.' In Australia, for example, the sheer size of the fields precludes transplanting, and highly accurate control of

China – *Near the town of Yangshuo, Shui Lian slides a pole through the flowering rice plants, spreading the pollen to fertilise the rice plants.*

GROWING 35

water depth reduces the possibility of seedlings being washed away or waterlogged before they have become established. Direct seeding, into either dry or wet fields, uses less water than transplanting, principally because it takes much less time to prepare the field in a flooded condition.

Once the seedlings are established and growing, there's plenty to keep the rice farmer busy, whether it's ensuring the water is maintained at the optimum level or keeping any encroaching weeds under control. Fighting pests and adding fertiliser are also regular procedures as the rice grows and matures.

India – Near Ghaghraghat in the northern state of Uttar Pradesh, a farmer uses two oxen to plough a recently harvested field.

Indonesia – A farmer ploughs and puddles his field near Rendang in the east of Bali. In tropical rice-growing areas, where there is no specific growing season, fields only lie fallow for a few weeks at the longest before the next crop is planted.

PREPARING THE FIELD

In Asia the stages of preparing a rice paddy for transplanting are essentially the same wherever you go. After the last harvest the rice stalks are either burnt off or cut down to be used as straw for animal fodder. Then the field is ploughed; in some areas small 'walk-behind' tractors are coming in to use, but in large areas of the developing world animals are still used to pull the plough.

Once the field is satisfactorily ploughed it is flooded and then ploughed again and puddled to make the field as level and the soil as uniformly muddy as possible. The more level the field, the less water will be required to flood it and the better the rice will grow. Ideally the crop should not be left high and dry nor, at the other extreme, should it be flooded too deeply. In more advanced rice-growing areas the field will be laser levelled using a plough which is automatically raised or lowered according to signals from a laser.

Vietnam – *In the Mekong Delta a farmer weeds the flooded field before transplanting.*

In the Balinese village of Iseh a group of villagers works at transplanting the seedlings. The terraces step down to a fast flowing river far below, then climb up the other side of the valley towards Gunung Agung, Bali's holy mountain. Hulking ominously to the north Agung erupted violently and disastrously in 1963.

INDONESIA

Transplanting in Bali

In most rice growing areas in third world Asia rice goes through a transplanting process. Seedlings are grown in a nursery field, either one field in a group or perhaps just a corner of a larger field. Only when the seedlings have become well established are they then transplanted to the larger field.

There are a number of advantages to this process. In the small nursery field the seedlings can grow to the stage where a sudden flood or heavy rainfall won't overwhelm them, they're big enough to face the outside world when they're transplanted to the larger field. Furthermore, they're certainly much bigger than any potential competitors; they've got a good head start on any weeds that might pop up beside them following the transplanting operation. This has the added benefit – ecologically and financially – that farmers don't need to rely too much on herbicides to keep weeds at bay.

The downside of the transplanting operation is that it's enormously labour intensive and time consuming. A Balinese family might own half a hectare to a whole hectare (2-1/2 acres) of rice paddies but the whole village group turns out for the transplanting. The only other time in the entire rice growing process that so much intensive labour is lavished upon the rice paddies is when harvest time rolls around.

In the village of Iseh the fresh young seedlings are uprooted from the nursery fields and tied into handy bundles, then trimmed to a uniform length against a sharp-edged sickle which has been stuck firmly into the mud. Already another worker is stirring up the nursery field, ready to plant that out as well. The bundles are tossed across the field edge to the main group, who work backwards across the flooded terrace, pushing the seedlings into the soft mud.

I want to know what it's like to start a rice plant on its way to the table so I pull my sandals off and park them with the line of footwear on the field edge. The mud oozes up between my toes as I step, gingerly, into the field. No explanation is necessary, a bundle of seedlings is tossed over to me and when I poke one into the smooth mud it's quickly explained that I should plant three or four – *'tiga atau empat'* – at a time.

In the village of Ceking a farmer ties up and trims bundles of nursery seedlings, ready for transplanting.

A truckload of rice seed is delivered to the airstrip and loaded on to the planter aircraft at Moulamein.

AUSTRALIA

Aerial planting in Moulamein

In many parts of Asia, growing rice involves transplanting, laboriously plugging nursery seedlings into the soil a handful at a time. The affluent Japanese have mechanised the transplanting operation with ingenious machines, loaded with lawn-turf-like mats of rice seedlings which are put out with metronomic efficiency and accuracy. It eases the hard labour of transplanting but it's still a time consuming business.

In Australia the seeding is done on an altogether different scale: by aircraft. The field is fertilised then flooded two weeks before sowing. Just 24 hours before the mid-October sowing, the rice seeds are 'pre-struck' and sown from the aircraft flying low over the flooded fields. Some of the sowing is done by tractor but 95% of Australia's rice crop starts this way, by air.

A planter aircraft skims over the flooded bays of an Australian rice farm at Moulamein in New South Wales.

Three Ifugao women transplant a Banaue rice paddy, one of these energetic riceworkers is in her mid-60s but the other two announce that they're about 75.

PHILIPPINES

Transplanting in Banaue

Ten minutes by road north-east of Banaue the aptly named Viewpoint looks out over a classic terrace view, hopscotching down the valley. Fight your way through the souvenir stands and suddenly time winds back and the rice planting activity carries on as it has done for over a thousand years.

Elena, Inggihi and Udayya are three cheerful Ifugao women busy transplanting the rice. They're not worried by the backbreaking work, they explain, because they've been working in the rice paddies since they were so high. We explore down the valley a little way, clambering down the steep, stone-walled terraces. The mountain air is a cool escape from the sea level heat of Manila. It's possible to tightrope walk along the terrace edges all the way back to Banaue, a three hour walk. Instead we walk back up to the Viewpoint, meeting our three elderly transplanters, their work finished, on the way. Richard is courteously ushered to the front of the line, youth must go first they explain.

Transplanting rice fields can be hard on the back but that doesn't seem to bother Udayya, an Ifugao woman in her mid-70s.

PROBLEMS

If everything goes well you plant your rice, look after it as it grows and harvest it when it's ready. But things don't always go right. The rains can come late and your crop can wither for lack of water. Or the rains can come too heavily and simply wash it away. Rats can come and steal the rice just as it's almost ready to harvest, rats will even swim out to deepwater rice. Then there are diseases and pests, blast which can cause the rice to wither and die or simply produce empty grains, leaf folder, stem borer and in India a disease known simply as gundhi, 'bad smell.' Snails can devastate a crop or birds can simply peck it off, grain by grain.

'We treated the rice but we were too late,' said Phool Mohammed, studying his rice field at Ghaghraghat in Uttar Pradesh. 'If we could have afforded to treat it earlier it might have been saved, but the leaf blast spread to become neck blast and now we have lost almost the entire crop.'

Around him the rice should have been ready to harvest but the panicles were toppling over and the rice grains were mainly empty. Blast is a fungal infection and it has clearly devastated the crop.

'Still it was even worse last year; we had floods which made the rice more susceptible to disease. Last year we lost everything. This year it was dry so we didn't think it would become infected again. Otherwise we would have treated it earlier.'

Top:
  India – Neck blast or panicle blast has caused the grains to wither and die.
Bottom:
  Philippines – Red snail cells look like raspberries attached to the stems of rice plants at Batad. When they hatch out they can devastate the rice crop.

India – A disconsolate Phool Mohammed examines his damaged rice crop.

Women's Work in Nepal

At a field at Lele, a village just outside the Kathmandu Valley, Kanssi (the name means youngest daughter) and her daughter Usha are clearing a drainage channel to try and remove the excess water from their field. The monsoon has continued late this season and the fields, which should be drying and ready to harvest, are still waterlogged.

Kanssi's family are Magars, a Nepalese ethnic group who are both farmers and soldiers. The largest proportion of the famed Gurkha army regiments are Magars. Her husband was one of three sons who each inherited three *rupani* of their father's nine *rupani* of flat land at the bottom of the valley. A *rupani* is just under 500 square metres so the father's original nine *rupani* would have been about half a hectare. Three rupani is not enough to live on so he works as a truck driver. He doesn't like rice farming in any case, Kanssi reports. Apart from the planting he leaves most of the work to her.

Kanssi and her daughter Usha work at clearing excess water from their flooded rice field.

HELPING THE CROP

Herbicides, pesticides and chemical fertilisers all require great care; too much is often worse than too little. The modern improved rice varieties usually require outside help in the form of fertilisers, but training farmers not to overdo the assistance is a key factor in raising yields. Farmers have become similarly dependent upon chemical assistance to keep weeds and pests at bay, but who wants to go back to the days when rats could consume the entire harvest?

Surprisingly, farmers in Australia are often much less dependent upon these scientific additives than their counterparts in the developing world. Rotating crops, a luxury often not available to an Asian farmer with his tiny patch of field, means that fertilisers are either not needed at all or needed only in small quantities. Combine crop rotation, which interrupts a pest's life cycle, with Australia's relatively remote and isolated rice growing areas, and the dangers of pests and diseases are also greatly reduced.

In Burma and Vietnam fertilisers and pesticides are in common use. Prior to World War II Burma was the world's

Australia – In a virtuoso display of low-altitude flying fertiliser is sprayed on to the flooded fields in New South Wales.

Vietnam – *In the country's Mekong Delta ricebowl a farmer sprays pesticide on his field.*

largest rice exporter, but the inept government's mismanagement saw rice exports decline from 1.7 million tonnes per year in 1962 to barely 200,000 tonnes by the late '80s. In recent years the long-term fall in production and exports has been reversed and exports are once again close to one million tonnes per year. Burma devotes 70% of its cultivated area and 40% of the labour force to rice growing, but primitive equipment and farming techniques, poor quality cultivars, ancient processing plants and a stunted marketing and trading system continues to limit rice-growing potential. The rice business suffered similar mismanagement in Vietnam until the economic reforms of the late '80s rapidly transformed the country from a rice importer to the world's second largest exporter.

Burma – *Near Pathein in the Irrawaddy Delta a farmer scatters fertiliser over a recently planted field.*

Following Page: *Indonesia* – *Villagers work their way through a rice field on the island of Lombok.*

IRRIGATION

Rice needs water to grow and apart from the tiny percentage of dryland hill rice, almost all of it is grown in flooded fields. That may be a picturesque series of paddies stepping down a hillside in Bali, China or the Philippines, with the water carefully organised to flow from field to field down the hill, or a flat field on the plains of Australia, Burma or India's Punjab where the water is pumped to the fields from irrigation channels and canals.

Organising the irrigation water for rice fields is often a politically charged situation. In Bali the *subak* organisation is effectively a rice growers' local government with responsibilities not only to ensure that the irrigation channels and waterways are maintained, but also to make sure that everybody gets their fair share of the precious fluid. It's often said that the best person to be in charge of the *subak* is the farmer at the bottom of the hill; he has a considerable incentive to ensure the water flows all the way down!

Supplying irrigation water is an equally tricky state of affairs in Australia where water shortages are common and concerns exist about the dangers of land degradation and soil salinity as a result of irrigation projects. Consequently, Australian rice farmers have to be very careful to select clay-type soils to minimise the amount of water that leeches down into the subsoil. Furthermore, rice crops are not repeated immediately; usually two or three years will pass before rice is planted again in the same fields. The price farmers have to pay for their irrigation water is equally problematic. The expected price for the crop has to be balanced against the price for irrigation water, while at the same time the water sellers are pondering just how high they can push the cost of water. Remarkably, rice was introduced in Australia in order to use more water, because in those days irrigation projects could supply more water than the fruit growers, who first moved into the region, were willing to buy. Today it is widely questioned whether growing rice on the world's driest continent is really a good idea.

China – In the Dragon's Backbone Rice Terraces water generally flows down from terrace to terrace but sometimes ingenious bamboo channels are used to carry the water longer distances.

Australia – *An expert eye studying this aerial view of a recently sown contour bay near Moulamein in New South Wales would notice that ducks had been pillaging the seed shoots.*

Harvesting

At harvest time, scale, technology, affluence and culture all blend to paint a different picture in different countries – just as they did when the rice was first planted.

In developing Asian countries the process is generally very small scale, unmechanised and neighbourly. The workers usually move through the field wielding sharp sickles and will later thresh the grains from the stalks by hand. It's an age-old technique, both simple and effective. Beam a Balinese villager down into a Chinese paddy field, put a sickle in his hand and he'd know instantly what to do and how to do it. Yet it's the family, the village and tradition that really comes into play at harvest time. In Bali the whole village will turn out in a labour that's partly neighbourly assistance and partly civic duty. Fail to appear and you can be fined. In China, big city industrialisation has depleted the permanent labour force, but come harvest time many young people return to their families for a few days of hard work to bring the rice in.

The picture changes with increasing affluence. In Thailand, world champion rice exporters, a local rent-a-harvester operator may turn up with a modern combine harvester but family and friends will still turn out to help. In Punjab, the get-ahead north Indian state and homeland of the turbaned Sikhs, mechanical equipment is again to the fore.

Japan and Australia provide a startling contrast to developing countries and to each other. In Japan the harvest is highly automated but very small scale. The combine harvester may be modern as tomorrow but it's tiny. Modern equipment and techniques has reduced Japanese rice farming to a hobby. Apart from a few weekends at planting and harvesting time, growing rice only requires the odd hour here and there. Rice farming has become a spare time activity; and one pursued by ageing enthusiasts.

Very soon Japan's famous insistence on rice self sufficiency and reluctance to import the grain may be stymied by a simple lack of rice farmers. Younger Japanese are proving increasingly reluctant to follow their parents on to the land. Contract farming for an increasingly absentee farm owning community is becoming the norm.

It's all different in Australia. Once again the equipment is state of the art, but here the scale is huge. Statistics eloquently illustrate the story. A hard-working farm couple can look after 300 hectares and count on bringing in 3000 or more tonnes of rice in a season. Their equivalent in Asia might have just a half a hectare and bring in only two or three tonnes each season, although admittedly the climate may permit double or even treble cropping.

Indonesia – At Ceking near the village of Ubud in Bali a rice farmer ties up a bundle of harvested rice.

Golden rice fields and artistically arrayed bundles of rice straw make a beautiful picture at the village of Nan Mu, close to the Guilin-Longsheng road.

Baskets full of recently threshed paddy rice.

CHINA

Harvest Time at Guangxi

The rice crop is brought in, tiny field at a time, in the Guangxi villages along the road to the regional centre of Longsheng. From this district most of the young people make an annual exodus to the noisy factories of frenzied Guangzhou (Canton) near Hong Kong, returning only at harvest time when labour is needed to bring the crop in. The roads are crowded with what looks like double-decker buses for dwarfs; in fact they're crammed with two levels of bunk beds for the long trip to the big city.

As each field is harvested at the village of Nan Mu, the rice straw is neatly tied up in conical bundles to be stored away as animal fodder for the winter. It's a scene of rural Arcadia but rice farming does no more than put food on the table, it no longer brings in enough money to support a family.

At the village of Nii Jiang 47-year-old Lii Jiao, her 18-year-old son Ji Yuan and a group of his friends are threshing the recently cut rice into two large wooden containers. It's a day off school for a teachers' festival and the family has 3-1/2 *mu*, about two-thirds of a hectare. There's no more land available in the area so what happens when the children grow up?

At the village of Nii Jiang, Lii Jiao winnows the recently harvested rice.

... and bags it.

In a scene that could be a century old rice straw is loaded on to a bullock cart.

BURMA

Harvest Time at Pathein

In a small village outside Pathein, four women and a man work across a beautiful golden-green rice field. It's a quintessentially Burmese scene: a clear, sunny day, formations of dragonflies hovering overhead, the workers in their traditional *longyis* and straw hats, the women with their faces powdered with *thanakha*. It's a good harvest they announce, they expect to bring in as many as 100 baskets per acre. Pathein, in the centre of the fertile Irrawaddy Delta region, has long been the centre of the country's most productive rice growing area. The workers are paid less than US$1 a day at the official exchange rate. Mya Thet Mon, one of the young women wielding a sickle so effectively on this small field has a special connection to the rice she harvests, her family own the field.

When you're harvesting rice with nothing more than a sickle it's hard to tell where the scene is set, in this case the typically Burmese straw hats give the location away.

The straw hat and thanakha face powder confirm that we're in Burma where Mya Thet Mon's family own the rice field she is working so diligently to harvest.

INDONESIA

Harvest Time at Ababi

Soon after dawn villagers from Ababi, just a km from the water palace at Tirta Gangga in eastern Bali, move on to the fields of I Made Geria. Terracing runs from the strip of forest that separates Ababi from Tirta Gangga right down to the road, and his holdings total 55 *ara*, an *ara* is 100 square metres or a 10th of a hectare. The land owner is joined by I Wayan Dauh, head of the local *subak*, and by more than 50 villagers, all members of the *subak*. The *subak* administers the most important constituent in Balinese rice growing – water. It's the *subak's* responsibility to ensure that water makes its way from the top-most right down to the bottom-most field in its area and every villager who owns as much as one *sawah*, a single rice field, must be a member of the local *subak* and take part in its activities.

The *subak* also supplies the communal tools of the rice growing trade, the racks on which the rice is threshed for example, and the labour. All *subak* members turn out to help harvest other members' crops and there's an element of carrot and stick in this co-operative labour. On one hand 10% of the harvest is distributed amongst the harvesters, on the other hand no shows have to pay about 60 to 80 cents for failing to lend a hand.

The harvest has been proceeding on I Made Geria's fields for several days and it will only take a few hours work this morning to finish bringing in the crop. Higher up the hill the rice stalks are rapidly cut, bundled and carried downhill to the large blue tarpaulin where they will be threshed. Long, angled bamboo racks are arranged at both ends of the tarpaulin and the threshers whack the rice stalk bundles on to these racks, sending the rice grains flying on to the tarpaulin.

It only takes a few hours work to complete the harvest, at which point the paperwork begins. The field owner carries an exercise book listing the villagers who should have turned out and calls roll so that the absent workers can be charged for their non-appearance. The *subak* head supervises the bagging of the rice, it has been a good harvest and the morning's work has yielded 22 bags, each weighing 40 to 45 kg, so there will be nearly 100 kg of rice to divide up between the workers. At this point in the proceedings Jero Mangku Diksa, the Ababi village priest, also turns up because 15% of the harvest will go to the temple. The owner's rice bags are now carted downhill to the roadside for a passing *bemo*, public minibus, to take them back to the village.

I Wayan Dauh, head of the local subak, the Balinese rice-growers co-operative, checks off the subak members' names in his workbook.

A villager rakes over discarded rice stalks.

Already the threshing racks and the blue tarpaulin are being hauled uphill to the next harvest. The owner, *subak* head and village priest squat under a shady tree to discuss the harvest. Ni Wayan Runi has already closed up shop and headed uphill. She had established a small shop in a *pondok*, the open-sided grass-roofed shelters which dot the rice fields, to sell snacks, hot coffee, *kreteks*, the Indonesian clove-flavoured cigarettes, and other necessities to the harvesters. Finally there's just a small girl and an old lady left in the field. As each bundle of rice stalks was threshed and tossed back over a thresher's shoulder another small contingent had raked over the discarded rice stalks, catching the few rice grains which hung on despite the threshing. These last two rice-workers are busy filling their own small sacks with these escapees.

The sickle is the standard tool for harvesting by hand right across Asia.

A group of farmworkers thresh the rice stalks against a metal cage.

INDIA

Harvest Time in Kaul

Near the Kaul Rice Research Station in the state of Haryana, India the centre has a dozen experimental fields where new crop varieties can be tried and tests conducted with new machinery. A group of laughing farmworkers thresh the rice on a simple metal cage after which the winnowing is done in an equally straightforward fashion but with a breeze provided by a large fan driven off a tractor engine.

Simply flapping a sheet generates enough of a breeze to winnow the rice in this field.

INDIA

Winnowing

All along the road between Lucknow and Ghaghraghat in Uttar Pradesh harvesting of early rice is underway. Because the monsoon rains arrive later than in Punjab the harvest also takes place later. Once the rice has been harvested and the grains threshed from the stalks the next stage in the process is winnowing, separating the grain from the chaff.

Letting the wind do the job by simply tossing grains and chaff into the air and watching the chaff blow away is a time-honoured process but how do you generate your own breeze if nature doesn't co-operate? In a field by the road the vicious looking blades of a hand-cranked metal fan does the job very efficiently. Further along we stop to watch another harvesting scene where a windbreak of stacked rice straw and a simple flapping sheet provides the necessary air current.

In a farmhouse in Ghaghraghat tossing grains and chaff into the air is all it takes to winnow the rice.

NEPAL

Harvest time in the Kathmandu Valley

The approaching new moon festival of Deepawali, also known as Tihar, is another reason to hurry the harvest. On that dark night candles and lamps are lit throughout the country as the enchanting Festival of Lights welcomes a visit by Lakshmi, the Goddess of Wealth. In Nepal, as in so much of Asia, rice is wealth so new rice will be offered to the goddess.

The traditional Nepalese rice grown in the valley is Marashi but it's been supplanted by the new Chinese variety known as Taichin. Nevertheless, the local rice is not to the taste of the valley's middle class who prefer the 'lighter' Manasuli from the lowland Terai region on the border with India.

'Valley rice is too heavy,' announces our guide, Adwait Pradhan. 'Eating it weighs on your stomach and you feel tired and lazy.'

At the eastern end of the Kathmandu Valley the first clear day after the monsoon immediately brings workers into the field to harvest the rice.

In the developing world threshing the rice is often done in the field and by the simplest means imaginable, simply whack it against a tarpaulin.

Sporting the traditional conical non bai tho hat, a farmer spreads the rice out over a blue tarpaulin to dry near the village of My Thuan.

VIETNAM

Harvest Time in the Mekong Delta

Along the main route from the Mekong River port of Cantho through Mytho to Ho Chi Minh City (Saigon) there's constant rice producing activity. The Mekong Delta, south of Ho Chi Minh City, is the rice bowl of Vietnam and in recent years it's been a very successful and very productive rice bowl. In 1986, 10 years after the final close of the Vietnam war, economic reforms known as *Doi Moi* or 'renovation' were commenced. Basically it meant more individual enterprise and less of the state running everything. The results were almost instantaneous, in just three years Vietnam changed from rice importer to rice exporter. The growth in exports continued steadily into the 1990s and today the country is one of the world's largest rice suppliers.

Despite the huge increase in production most of the rice is produced in a decidedly low-tech fashion. An estimated 70% of the population is involved in growing rice, but it's the Mekong Delta where most rice is grown – 50% of the national total – and this same region supplies almost all the exports.

At the village of Cai Lay, as the last bags of threshed rice are tied up, the field is already being burnt off while the threshing machine is carried away to the next field.

Rice is bagged on the harvester while two of the workers carry a bag of rice up to the roadside to be loaded up and transported back to the village.

THAILAND

Harvest Time in Pathum Thani

The Thais are very proud of their fiery cuisine, which in recent years has become popular around the world. Fragrant rice, of course, is an essential ingredient. Thailand is not one of the world's gigantic producers and consumers like China, India or Indonesia but when it comes to export it's right up there at the top. Thai rice finds its way to the tables in over 120 countries around the world but it's the high quality Thai Jasmine rice which is best known.

Near Pathum Thani, not far north of Bangkok, Somporn Tuammai's 1-1/2 hectare field is being harvested. About a quarter of the country's rice crop is brought in by machine and here a local harvester hire company provides the machine and its driver for about 2000 baht (US$50) a hectare. Mechanised or not it's a typical friendly Thai scene with lots of people working and an equally larger contingent simply watching.

Apart from Somchai, the driver, and Somporn, who supervises the whole show, there are three other women and one man riding on the Isuzu harvester to bag the rice as it comes aboard. Two of Somporn's sons and some neighbours turn up with those walk-behind tractors which are such a part of the agricultural scene in many Asian countries. They'll transport the grain back to their village.

There's a lot of smiling, joking and laughing, they're clearly a fun loving bunch of rice farmers but also a hard working bunch. In just 10 days time they'll be planting the field again, and with the modern, fast maturing variety they grow they expect to harvest two to three crops a year.

'This is a good field,' announces Somporn's cousin.

Rice field owner Somporn Tuammai has a smile on her face as she surveys the harvest scene.

*The combine harvests just three strips of a rice at a time, about a metre wide but the edges
and corners of the field are still cut by hand. The rice is a dwarf variety, only reaching waist high.*

JAPAN

Harvest Time in Tsukuba

Many Japanese rice growers are, like the Yokayamas, 'weekend farmers', although unlike western 'hobby farmers' it's simply that rice growing has retreated to becoming a weekend activity, it's not something taken up for weekend amusement. Altogether their rice fields in the Tsukuba area north of Tokyo, total about 60 *are* or 0.6 hectares which makes them one of the smallest rice farmers in the district. That's certainly not enough to make rice farming a full time activity and Mr Yokayama points out that he's dressed more like a garage mechanic than a farmer, confirmed by the sign announcing Yokayama Car Repair on the side of his truck. He looks like he belongs under a Toyota rather than in a rice field.

Even as a weekend sideline rice farming is only a break-even operation for the Yokayamas and what's going to happen after this generation? Their three children help with the planting but all have full time jobs and no interest in continuing the family rice growing tradition. Fortunately it's not a job requiring much labour. Everybody helps to plant the rice but after that it only requires a little fertilising and weeding and a couple of weekends of harvesting for the husband and wife team.

It's an early September harvest due to a late spell of hot weather, but Mr Yokayama is concerned that the late heat wave has left the rice too dry and he's unsure of the quality until the rice is milled. First grade rice fetches US$2.50 to US$3 a kg when sold through the farmers' cooperative to the government but if this is second, or even third grade, which only fetches US$1.50 a kg, then he may try to sell it privately.

Another farmer drives by in one of the tiny pickup trucks which seem to be the badge of authenticity for a local rice grower.

'It must be *atsui* (hot) out there,' he shouts to the Yokayamas, the sun is indeed beating down.

'No, it's *iyambei* (comfortable),' replies Mrs Yokayama, using the old fashioned word meaning well balanced.

To an American or Australian rice farmer the Yokayamas' combine would be a toy, a cute US$17,000 Iseki 1400. Some farmers rent a combine or pay someone else to harvest for them but most, like the Yokayamas, own their own. The combine cuts the rice and packs it into neat little 20 kg zip bags hung on the side of the machine.

JAPAN

A Full Time Farmer in Tsukuba

Not far away Toshio Ogura is another side of Japan's small scale rice farming picture. He's a full time farmer although he only owns a tiny 10 *are*, one tenth of a hectare, himself. He rents other fields and also harvests rice for other farmers. His combine harvester is a much larger device though still toylike by American or Australian standards. It's a US$50,000 Kubota SR50 Skyroad Pro which cuts no less than four rows at a time and transfers the threshed rice straight into the G17 Grain Tank on the truck which his wife drives. He can lift the tank off the truck with his Kato 'My Crane'. He's proud of his equipment but the corners of the field still have to be trimmed by hand.

Despite the larger scale of his operation, Toshio Ogura faces the same worries as his neighbours. He is in his early 50s and reckons he's the youngest farmer in the area. His two daughters don't work on the farm, they both have other jobs, and there are no successors. The farmers he works for are now in their 70s and too old to work their fields.

The field is full of herons, *sagi*, waiting for grasshoppers to pop up out of the field as the combine approaches. As if at a signal they suddenly all fly up and move on to another field. Like the Yokayamas, Toshio is worried that the rice won't be first grade because of the late season hot spell. Farmers who were late in draining their fields may fare better.

Full time farmer Toshio Ogura surveys his fields.

Transferring grain from the combine to a grain tank on the truck.

Harvest Time in Yasato

Even the larger Japanese farms would seem very small scale to an American or Australian farmer so the small ones are really miniature. In the village of Yasato, on the slopes of Mt Tsukuba, harvested rice stalks are put out on racks to dry in the sun. No herbicides or chemicals are used on this crop which is also weeded by hand. The neat little lawnmower-sized harvester spits out neatly tied bundles, ready to be hung over the drying racks.

The harvested rice stalks are laid out on racks in the sun to dry. 'It tastes better when sun dried,' Tsuneo tells us. 'And it dries quickly when the weather is like this, atsui (hot).'

The technology can get very small indeed, Tsuneo Imahashi walks behind his tiny harvester which handles just two rows of rice at a time, neatly bundling up the rice stalks and laying them down to await threshing.

Kerry Lowing pulls the tractor or 'chaser' with its 'storage hopper' up beside the header to transfer the harvested rice across.

AUSTRALIA

Harvest Time in Moulamein

Rice may seem a strange sight in Australia but in other ways it's a typically Australian scene. Roos bound away across the road as we approach the field, cockatoos wheel overhead and there are lots of snakes and frogs in the fields. Rice has been grown commercially in Australia for about 50 years, almost all of it in the southern part of the state of New South Wales using irrigation water from the Murray and Murrumbidgee Rivers.

Kerry and Nick Lowing have about 200 hectares in rice in the summer and about 600 hectares in wheat in winter at their Burrindi farm, near Moulamein. They'd grow perhaps 50% more rice if there was enough water available.

Kerry and Nick are in radio communication between his combine harvester – known as a 'header' even though correctly only the front part is a header – and her tractor, the 'chaser', towing the 'storage hopper' or 'chase bin' or 'auger bin'. When the storage area on the header is full they drive alongside each other and pump the rice across. When the chase bin is full Kerry drives over and transfers it to the much bigger 'mother bin' which holds about 90 tonnes. In turn trucks come by and take about 25 tonnes at a time to the mill.

They can't start harvesting until about 9.30 am when the sun gets up and dries off the dew. They have to stop by early evening as the moisture content in the rice starts to rise as the temperature drops. The mill closes earlier, at 4 pm, but that's not a problem as they can continue to fill their mother bin. Working this schedule they can harvest 25 to 30 hectares a day, they'll bring in the whole crop in a week. It takes just two of them to run the farm.

It's not been a bad year, a cool but consistent summer and fairly humid, which is good. A large proportion of Australian rice is exported and Papua New Guinea is the main market although some also goes to Japan. Koshihikari, Koshi to the Aussie farmers, is specifically grown for Japanese preferences. It's not a high yield variety; they get seven to eight tonnes a hectare where high yield rices will produce over 10 tonnes a hectare. The Lowing's John Deere header costs about US$200,000, it's a 9600 combine with a 6022R header. Nick's a firm believer in having up-to-date machinery: ' you simply cannot afford breakdowns.' The header on this harvester just pulls the grains off, doesn't cut the stalks.

He invests in a new header about every four years and does some contract harvesting although 70% of farms have their own equipment. Contract harvesting of wheat is done by the hour, rice by the ton. 'Rice is more abrasive,' Nick says, 'it wears headers out faster.'

High-Tech Farming

Inside the header it's all very scientific with a computer readout in the top corner of the cab which Nick constantly surveys. The most important figure is the bottom one, indicating the moisture content of the rice. 'Ideally it should be 18 to 24%,' says Nick, 'above that the rice will be too moist to dry properly, below that it will be too dry and it'll crack when it's milled.'

Higher up the screen are yield figures, showing how many tonnes per acre are being harvested. The Lowing's fields are laser-levelled to make the most efficient use of the irrigation water. The field slopes down by perhaps 8cm over 100 metres but in that distance the slope will not vary by more than about 5mm. These remarkably precise levels are achieved by a great deal of ploughing and as a result the depth of topsoil varies. The rice yield varies with depth and quality of topsoil so the readout dips down to as low as two tonnes, then jumps up to as high as six.

'We'll upgrade the computer system with the next header,' Nick reports. 'It'll be linked to a GPS (global positioning system) and the computer will record those yield variations. Then when we're spreading fertiliser before we sow the next crop the computer will automatically adjust the fertiliser level to compensate for the areas where the topsoil is thin and the yield is low.'

As the header moves through the field a computer readout gives a continuous report on the yield in tonnes per acre and the moisture content of the harvested rice.

Rice pours into the mother bin

Their equipment arrayed behind them Nick and Kerry Lowing check their rice crop.

Processing

Each stage in the rice story from planting the seed to its final appearance as an about-to-be-consumed bowlful on the table follows essentially the same pattern. Whether it's pushed into the soil as a small clump of seedlings in the time-worn transplanting process or broadcast from an aircraft, the planting process is simply a matter of getting the plant into the earth. At maturity the rice has to be harvested, a straight forward matter of transferring the grain from the field to storage, although the process can be totally mechanised or involve nothing more sophisticated than a sharp sickle and a strong arm.

It's a similar story at the processing stage. Removing the protective husk and milling away the individual grain's coating of bran to produce the final product, white rice, can be done with a sophisticated, computer-controlled mill or a simple wooden mortar. The end result is the same.

The rice trail led us to both extremes of the rice processing story. In the Philippines, close to the age-old rice terraces of Banaue, we watched a villager turn rice grains, still on the stalk, into polished white rice using nothing more

sophisticated than a hollowed out wooden mortar, a hefty log as a pestle and some serious muscle power. In Pathein, in Burma's fertile Irrawaddy River delta, we watched as a boat-load of rice was unloaded, sack by sack, and processed through a mill which looked as if it could have featured in a tale by Rudyard Kipling. At the rice mill we visited at Cai Rang in Vietnam's Mekong River delta, the rice also arrived by boat, although the mill was certainly a little less ancient. In Thailand, Asia's other major rice exporter, our travels took us to a much larger and much more modern riverside mill in the former capital of Ayuthaya, just north of Bangkok.

Rice processing in Japan and Australia proved a contrast to the other regions on our itinerary and to each other. The processes and equipment were totally modern but, as with the other stages of the rice story, the scale of operations and equipment was tiny in Japan, huge in Australia.

Bangladesh – At a rice mill in Latifpur, north-west of Dhaka, rice husks are burnt to boil the water to make the steam to parboil the rice.

INDONESIA

Milling in Bebandem

In the east of Bali a small local rice mill operates at Bebandem. Rice is spread outside on concrete platforms to dry in the sun. Inside, bags of rice are fed into the mill and dehusked. The husks are blown out back where workers sift through the cast offs, winnowing them to let the wind carry off the husks, leaving rice kernels which have escaped the mill. Meanwhile the brown rice kernels are milled and the bran raked out to be used as animal feed. The milled rice is bagged in weighty 50 kg sacks and the rice owners, queuing up with their sacks of rice, pay in kind.; the mill takes 5% of the milled rice as payment for the service.

The rice husks are blown out the back of the mill where a woman winnows out the remaining grains which have managed to escape with the husks.

Drying Rice in Bali

Recently harvested rice usually has to be dried before it can milled and although that process is routinely done with hot air driers in the developed world, simple sunshine often does the job in developing countries. Many rice mills, such as the one at Bebandem in Bali, will have an adjacent flat area, often looking like an off-duty tennis or basketball court, where the rice is simply spread out to dry. The local road often provides an equally suitable drying platform and visitors soon become accustomed to driving nonchalantly across a village's rice harvest.

A Balinese back road makes a convenient drying place. The odd car or motorcycle driving over the unhusked rice doesn't do any damage although visiting chickens, like the one having a look at this rice, can be more of a problem.

On the river banks near Tha Yet Kohn village rice is threshed by simply walking cows across it.

The production processes are always straightforward in Burma. A farming couple winnow the threshed rice by spilling it off a tray suspended from a tripod frame. The farmer stands on an oil drum to get some height and the wind does the rest.

BURMA

The Antique Rice Mills in Pathein

Pathein, about 190km to the west of Rangoon, is the rice bowl of Burma. Much of the rice arriving at the town's mills comes by river, and at Kyauk Ngu, across the river from the centre of the town, a fleet of boats moor outside the government's 100-tonne mill. The rice is loaded in bulk and carried ashore a basket at a time. A sign outside the mill announces that their target for the year is to process 23 *lakh* (2.3 million) bags of rice. Stylish teak rowboat-ferries shuttle back and forth, the oarsmen standing up to row with a curious cross-oared technique, all of them sporting the hats without which no Burman looks fully dressed and exuding the calm, dignified look which is so much part of their country.

U Myo Wai - the mill owner

A few km upriver a boat is unloading 150 tonnes of bulk rice at the riverbank wharf of U Myo Wai's larger rice mill. Carrying empty rice sacks, an almost entirely female labour force trips delicately up a wet and springy plank onto the boat. Men tip two baskets of rice into the sacks and swing them up onto the womens' heads. Balancing nearly 50kg (100lbs) of rice on their heads the women stroll back down the plank with equal aplomb and tip their load onto the mill floor.

The mill, the largest in the area, is housed in two buildings; one dates from 1917 and one from 1948 but both look equally ancient. U Myo Wai has operated the mill for 35 years. The newer building was re-equipped in 1992 with World Bank help but the graders, the modern Thai-manufactured milling machines and all the other equipment are driven by an amazing number of belts, whirring round in all directions like something out of a Dickens novel of Victorian industrial England.

The mill floor shakes beneath the rattling, whining, sighing collection of Heath Robinson machinery. A spider web of big and small, short and long drive belts link this cacophony of equipment to a century-old British 125 horsepower, 90rpm, Marshall & Son steam engine. Ancient looking or not the mill can output 200 tonnes of milled rice a day and the milled-off rice husks are burnt to produce the steam which powers it.

Balancing sacks of rice on their head the female labour force unloads the boatload of rice entirely by hand.

The equipment may be antique but the mill certainly turns out the rice.

JAPAN

Rice Mills in Japan

Obata Rice Centre No 3, the Farmer's Corporation rice mill in Obata, is typically Japanese: shiny, neat, tidy, high tech – and tiny. It is also typical of the Japanese rice business in one other way: its workers are old. The seven men who run the centre – two to operate the centre's combines, two to drive its trucks and three to operate the machinery at the mill – have an average age of more than 70! Today's 50-year-olds just don't have the stamina, endurance and patience to be farmers, they report.

The corporation has 140 members in all, but most of them no longer farm; they rent their fields out and the centre harvests, dries, mills and bags the rice. There's a constant coming and going of the toy-sized pickup trucks used by all the local farmers bringing in bulk rice at one end while larger trucks depart with sealed 30kg paper sacks of rice at the other.

In contrast the Hojoh Country Elevator is a more commercial operation handling large quantities of rice in a drying, milling, packaging and storing operation. From the milling machine the milled rice is automatically dispensed into large paper sacks although the mill also produces smaller

At the Obata Rice Centre, Takao Tomita supervises the milled rice emerging at the end of the process and stacks the sealed bags of rice on pallets.

5kg bags for sale in supermarkets. The sacks of rice are numbered and stamped to confirm the quality and weight before they are palletised and transferred to the refrigerated storage area. The refrigerated warehouse stores over 1000 tonnes of rice.

Masaki Sakayori operates the control room in the Hojoh Country Elevator which processes the rice as it is dried, milled and packed.
Every 15 minutes he takes a sample to test for moisture content and other properties.

Bran, milled off the rice grains to produce the finished white rice, is raked into a heap in the mill, it will be processed for animal feed.

VIETNAM

A Rice Mill in Cai Rang

About 6km from the regional centre of Cantho on the Mekong Delta, Cai Rang is famous for its floating morning market. It's an important rice-producing area and there are 30 or 40 rice mills along the riverside. Te Van Hoa's mill is a medium size operation, capable of processing over 30 tonnes on a good day although overall it averages about 20. The owner, a rotund, genial gentleman clad in shorts and a Central Park New York T-shirt, is out on his wharf welding up some metal frames. He's had the mill for about eight years, not long after private enterprise got the nod from the government.

A boat arrives with 30 tonnes of rice on board, from a village about 35km from Cai Rang. Alternately supervising the unloading and wandering inside to watch the rice being processed, Nguyen Van Huong is a handsome man, 35 years old and slightly greying. He has two hectares of rice fields himself but for five years he's been working as a middleman, buying up rice from other farmers in his village, shipping it to Cai Rang and selling it to rice mills like Te Van Hoa's. The 30 tonnes on the boat will fetch about 50 million dong, about US$2500.

A web of belts spin, vibrate, rotate or, in this case, jiggle various bits of machinery.

In the mill there are about 40 young workers stacking the incoming bags and processing the rice. They average about 50,000 dong for an eight hour day, about US$3. A diesel engine at the back of the plant clatters steadily away, powering a web of belts which persuade various pieces of equipment to spin, vibrate, rotate or blow. At the front of the plant large, circular driers handle rice which is too wet to process immediately.

The unhusked rice is fed into a complex vibrating device which removes the husks. Nothing goes to waste; these rice husks are whisked out to a room at the front where workers bag them and load them on another wooden boat. They're sold to be used as fuel and once burnt they're used as fertiliser. The husked brown rice goes into the circular grinders which mill off the bran to produce polished white rice. The bran is, of course, the most nutritious part of the rice but it's not wasted either, ending up as chicken or fish feed.

Workers bag them and load them on another wooden boat. They're sold to be used as fuel and when burnt they're used as fertiliser. The husked brown rice goes into the circular grinders which mill off the bran to produce polished white rice. The bran is, of course, the most nutritious part of the rice but it's not wasted either, ending up as chicken or fish feed.

The mill owner's shy daughter Tu sits at the entrance to the mill, entering the details of incoming bags of rice.

A man with something to smile about, mill owner Te Van Hoa started his operation soon after the government decided to allow private enterprise in the rice business.

There's little mechanisation of the unloading process, the rice is heaved off a bag at a time.

At the Baba Rice Mill a group of colourfully dressed women rake, sweep and spread the parboiled rice across the concrete drying area. It takes a couple of good sunny days to harden the rice although breaking open a husk reveals that it's still quite soft inside.

BANGLADESH

Preparing Parboiled Rice

Latifpur is a town in the Andarkaliakur Thana, north-west of Dhaka, with numerous rice mills beside the river and along the road through the town. Bangladeshis, like the people of southern India, prefer their rice parboiled, a process which improves the taste (according to the locals) and makes it better for digestion but also reduces the percentage of broken rice.

Before it is parboiled the unhusked rice is dumped in large concrete tubs and soaked for 24 hours. This paddy rice is then carried in baskets and poured in the top of the vertical steel cylinders of the parboiler. Steam, generated by burning rice husks, passes through the cylinders for only five minutes before the rice is dropped out the bottom and carried away in baskets to be dried.

The women, who are paid about US$1 a day, go about their work with spirit and verve, dancing in a conga line as they spread the rice out to dry.

Cheerful, forward, fearless and quite assured.

PHILIPPINES

Husking & Milling by Hand

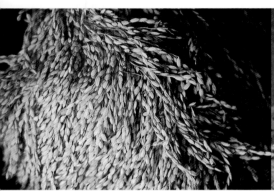

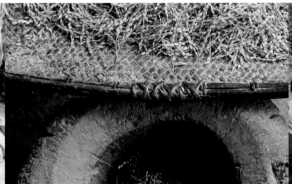

1. In the small village of Hacupon, near Banaue in north Luzon, the journey to the table starts with panicles of rice still on their stalks.

2. Kahlangan, a local rice farmer, threshes the rice, stripping the grains off the panicles.

3. The threshed rice grains are still in their husks.

4. The grains are tipped into a wooden mortar and Kahlangan and Dennis pound the rice using logs as pestles.

8. Many of the threshed rice grains are still in their husks.

9. The winnowed grains are poured back into the mortar.

10. Again the rice is pounded, removing any husks that escaped the first round but also scouring off the bran which coats the rice. This polishing operation converts brown rice to white.

11. The grains are now a mix of polished, semi-polished and unpolished.

5. The dehusked grains are poured out of the mortar onto a woven winnowing tray.

6. Grains and chaff, the husked rice is a mixture of grains and the broken rice husks.

7. Kahlangan winnows the rice by tossing it up in the air and letting the breeze whisk away the chaff.

12. Another winnowing operation separates the bran and the polished grains. The first winnowing operation simply dumped the husks on to the ground where they were quickly pecked up by chickens. The valuable bran, however, is caught in a second basket.

13. Aggin picks out any remaining unhusked grains as well as tiny stones and other debris by hand.

14. The end result is polished white rice.

AUSTRALIA

Processing & Storing

Rice has been grown in the Riverina district of the state of New South Wales since 1924. Starting with eight farmers around the towns of Leeton and Griffith in the Murrumbidgee Irrigation Area it has grown to become one of the key components of the local rural economy with 2000 farms engaged in rice growing. Most of Australia's more than one million tonnes annual production comes from this area, from where it is exported to more than 50 countries around the world.

Despite careful attention to the whole process, growing rice in Australia remains a very controversial subject. Farmers are only allowed to grow rice on 'heavy clay' soils which minimise the water loss and even then only 30% of the 'approved' area can be used for rice growing. There are strict limits on water usage and rice types particularly adapted to Australian growing conditions are used. Medium grain japonica varieties suited to temperate rather than tropical climates constitute 80% of the Australian rice crop. Crops are always rotated so rice is not grown on the same fields every season. Usually a wheat crop is planted immediately after the rice is harvested to utilise the remaining moisture in the soil.

From the farms rice is trucked to regional storage depots from where it continues to one of the region's rice mills like the one at Leeton. From there the rice heads out around the world, 85% of Australian rice is exported.

As each truckload of rice comes in to the Moulamein storage facility it is first weighed and then Neil Whitfield climbs up on the top and uses his probe to take a sample. Six samples have to be taken for every 10 tonnes of rice and there is approximately 26 tonnes of rice in each truck.

Rice pours into the gigantic storage facilities at Moulamein, one of more than a dozen in the region.

Paul Maytom inspects a sample of rice at the Leeton mill.

Trading

The world's rice production may be huge – over 500 million tonnes a year – but the amount of rice traded is, in comparison, minuscule. The big two in the rice export business are long established rice growing champion Thailand and recent new contender Vietnam, but as an export crop rice is absolutely nowhere; only a very small proportion is traded internationally. The USA and Australia, no big deal in the rice-production league, are nevertheless right up there when it comes to rice exports. Total worldwide rice exports amount to only about 16 million tonnes a year, not much over 3% of worldwide rice production.

Most rice is eaten at home; in fact well over 90% of the world's rice production is consumed in its country of origin, and in developing countries a very high proportion of the rice is eaten by the very same people who grow it. The two million tonnes of rice the US exports each year may give it the bronze medal for third largest rice exporter, but there are plenty of tiny rice-consuming countries which get through that amount of rice in a year. The Chinese devour twice that amount of rice every week of the year.

The Green Revolution may have temporarily halted fears of world starvation but continuing population growth is likely to once again put pressure on rice production. Exactly what the population of Asia will grow to is an open question, some countries have successfully put a brake on their growth but even China, home of the one-child-family, may have trouble maintaining that policy with increased prosperity. It's estimated that by 2025 the annual demand for rice in Asia will grow from the present 500 million tonnes to more than 800 million tonnes.

Thailand, Vietnam, the USA and Australia may be the world's biggest rice exporters over the long term but China, India and Pakistan sometimes pop into their ranks. When you grow so much rice it doesn't take very much extra to add up to a large number of tonnes. Surprisingly the countries of Europe are both big rice importers (2 million tonnes a year) and big exporters (1 million tonnes a year). Even countries like the USA and Australia, where a very large proportion of the production goes to export, still manage to import rice every year, usually fine regional varieties to satisfy rice gourmets.

India – *A variety of rice types on display in the market in old Delhi.*

SELLING RICE

Beautiful in the field when it's growing, rice is equally
attractive when displayed in markets anywhere in the world.
It's surprising how many different varieties of rice can be
found in even the smallest and most remote centres in Asia.

Vietnam – Papaya fruit is used to attract flies away from
the rice at the Cantho market.

Burma – A rice shop in Pathein, in the heart of the Irrawaddy Delta rice
growing region, offers several types of black rice.

Burma – The Burmese consume more rice per capita than any other
country, the Pathein shop also has a selection of white rice types.

Vietnam – Economic liberalisation spurred Vietnamese rice growers
to quickly become one of the world's largest rice exporters.
These varieties are also in the Cantho market.

Philippines – These varieties in Quiapo Market in the busy metropolis of
Manila sell for 17 to 22 pesos a kg, around 50 cents.

₱18.00/kg
TALAGA WALANG AMOY
REG. MILLED RICE
VARIETY: Maalsa, Masarap
GRADE NO. 1
₱23.00/kg
WELL MILLED RICE
VARIETY: Long Grain
PREMIUM GRADE
₱20.00/kg
FANCY RICE
SINANDOMING
₱22.00/kg
WELL MILLED RICE
VARIETY R-60
PREMIUM
20.00/kg
FANCY RICE
₱17.00/kg
REG. MILLED RICE
VARIETY: LAON Maalsa
GRADE NO. 1
₱18.00/kg
REG. MILLED RICE
VARIETY: Maalsa, Masarap
GRADE NO. I
BEAR

The dock workers aren't the only ones skimming off their share of the rice, these young boys were busy taking their own delivery underneath the truck.

Tally sticks record how many sacks of rice have been moved from ship to truck.

BURMA

Riverside Rice Theft

Rice arrives in Rangoon on the lurid brown, swift flowing Rangoon River by one of the admirably tatty fleet of ships that work the country's rivers. The *Zinbyun* ties up at the wharf and a continuous flow of coolies descend into the hold, sling 50kg jute sacks of rice on to their shoulders and march off down the pier to a waiting truck.

As they leave the ship they're handed a 'tally stick' for each sack, About a cm square by 25cm long these counters were an invention imported from India during the British colonial era. They come in different colours, for different types of rice or other produce. At the other end of the pier the tally sticks are handed in as the sacks are loaded on the truck. There's quite a collection of clerical workers to handle the collection of the tally sticks and mark them down in record books. The seller, the buyer and the rice wholesalers' association all have somebody on hand to record the operation.

Unloading the rice sacks is done manually; there's no mechanical equipment involved at all, and if that seems to have its costs there's another factor making this a far from totally efficient system. The labourers, coolies as they're known, are paid by the ship. Of course the bigger the ship

the more coolies are involved but typically the fee per ship is less than a dollar. A coolie may get through five ships in a day, giving him an income of about US$3 to US$4, but most coolies seem to skim off an extra little percentage for their work.

No sooner is a bag of rice slung on a coolie's shoulder than a small blade or pointed stick is used to cut a hole in the bag and as the bag is marched down the pier a steady trickle of rice spills out of the bag and into the worker's shirt pocket. In some cases this stolen rice is even fed down inside the shirt to a waiting bag hidden in the coolie's *longyi*, the all purpose sarong-like garment worn by men or women in Burma.

This organised theft is so blatant it's impossible for it not to be noticed. Every coolie marches down the wharf with a steady stream of rice spilling out of the sack and into a pocket, a sack or just the front fold of his *longyi*. At the end of the pier there's a security guard – he'd be a nightclub bouncer in the west – whose job seems to be to confront the more excessive skimming operations. In fact, all he seems to do is get them to hand over a part of their rice off-take. Perhaps he's just one of a series of percentage takers who also take a cut down the line!

Tally sticks in hand dock labourers or 'coolies' unload sacks of rice at the riverside docks in Rangoon.

U Myint Kyu, office superintendent at the Rice Wholesalers Market, checks his books in the market's shed-like home near the Rangoon River.

BURMA

Rice Traders in Rangoon

A cavernous old warehouse on the Rangoon docks houses the Rice Wholesalers Market, where wholesalers buy in bulk from the rice mills. It operates from 7 to 10 am, seven days a week and 5000 to 10,000 bags are traded each morning. Inside there are lots of tables and chairs where the buyers and sellers sit, pushing little piles of rice back and forth, inspecting it, picking it up and sniffing it, sifting through it, piling it up again, bargaining, making calls on their mobile phone followed by arcane mathematics on their calculators and scribbles in their notebooks. Finally the deal is made and lots of cash, since the biggest bill is not much more than a US dollar, changes hands. There are 600 licensed rice traders who can freely enter the building; others must pay a small entry fee of less than 20 cents.

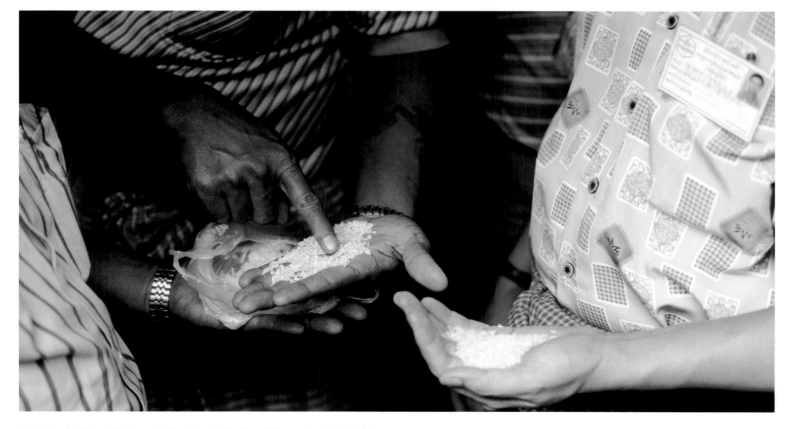

Rice dealers check the quality of a sample of rice, throughout the rice world the experts could make instant assessments of quality and moisture level without any specialised testing equipment.

Puffing on a king-size cheroot and with thanakha, the yellow sandalwood-like paste worn by many Burmese women as a combination skin conditioner, sunblock and makeup, on her face, cheery Daw Thein Win sifts rice.

Rice Wholesalers in Rangoon

The wholesale rice business is carried on at a number of locations around Rangoon. The Bayint Naung Bazaar is a collection of bare and dusty old warehouses where the Myanmar Rice & Paddy Wholesalers' Association can be found. On an average day five *lakh* (as in neighbouring India large numbers are counted in *lakh*, 100,000) 50kg bags of rice change hands in these warehouses. In the busy season that figure can rise to a million bags a day.

The rice and general trading business of U Nyein and Daw Ohn Kyin operates under the brand name of OK. It's one of the larger of the 200 or so rice traders at the centre, and in their warehouse are 10,000 sacks of rice of all sorts of grades. They're the middlemen, the wholesalers who buy from the rice traders who bring rice in from rice-milling operations around the delta. When business is slow less then 100 bags a day may move out from the warehouse to the markets and bazaars of Rangoon, but when it's busy they can shift up to 1000 bags in a day.

Daw Cho, daughter-in-law of the owner, sits at her desk flanked by a bank of telephones, each in a wooden box, for taking orders. Spread across the table beside her are small sample baskets of the 20 different kinds and qualities of rice available.

BANGLADESH

Rice Wholesalers in Dhaka

Just to the west of the Sadarghat, the popular boat landing on the Buriganga River in central Dhaka, is Maulavi Bazaar where several streets are dedicated to the wholesale rice business. Each small enterprise presents an almost identical picture: arrayed across the front of a platform are the samples, 10, 15, 20, sometimes as many as 40 small mounds of various varieties of rice. At the back of the platform the business proprietor, and perhaps an assistant or two, squats cross-legged behind a small wooden desk.

Arrayed in front of him are his ledgers, invoices and, inevitably, electronic calculator. Stacked down one side of the enclosure and in great floor-to-ceiling mountains at the back are the stock-in-trade, 85kg (nearly 200lb) sacks of rice. Along the street, between the rice dealers, are cafes, restaurants and small shops.

Outside it's a scene of typical Bangladeshi bedlam. At times the city seems to be one huge anarchic traffic jam where the same play constantly re-enacts itself, whether it's in a tiny one-rickshaw-wide back alley or an eight-lane highway. In Dhaka, if there's room for just one rickshaw it's inevitable that two will try and dispute the limited space. If there's room for four cars abreast you can be certain some chancer will try to start a fifth, then a sixth, then a seventh lane. Maulavi

A bag of rice is tipped off the weighing scales in the rice warehouse.

Bazaar is no different, there's a constant clamour and push, a cacophony of beeping horns and ringing bells as rickshaws and cars jockey for position while avoiding the trucks that force their way through the swarm of smaller vehicles.

In the middle of this confusion there's an equally constant arrival and departure of rice. Trucks pull up and deliver hundreds of sacks of rice, purchases are made and sacks are hauled away on grossly overloaded *vans* or on rickshaws and in pickup trucks. *Vans* are bicycle delivery utilities which, with much pushing and hauling, can drag off as many as 10 sacks at a time. A simple bicycle rickshaw can manage two or three, perhaps one less if there's a passenger on board as well. The river itself is only a few steps away and labourers haul barrow loads of rice sacks down to waiting boats, or collect incoming rice sacks heading to the wholesale dealers.

The activity is even more frenetic on the Buriganga River ghat. On the subcontinent a ghat is a riverside dock or quay although in this case it's just a muddy slope running down to the water. Life on this stretch of Dhaka's principal river is no different from on the road: colourful chaos reigns supreme and it's easy to pass an hour or two simply watching life unfold from the riverbank although renting a boat and getting out amongst the river life is even more interesting.

A van load of rice sacks is pushed and hauled away.

On the riverbank a melee of boats are being loaded with sacks of rice. A constant procession of labourers, hooks in hand, huge sacks of rice balanced precariously on their heads, trot down from the street and are ticked off by a bureaucracy of officials lined up at a table by the entrance to the ghat. They topple their sacks of rice onto the boats where more labourers arrange and stack them aboard. As usual there seems to be absolutely no order in the confusion. The boats are nosed in to the riverbank in wild disarray while amongst them a couple of men nonchalantly wash their hair and clean their teeth in the muddy water. The rice loaders have to clamber over riverside debris, tightrope walk across half submerged planks and scramble from boat to boat to deliver the goods.

Meanwhile there's an equally constant flow in the other direction. As fast as rice is loaded on to outgoing boats more rice is unloaded from incoming ones. A boat noses in with a load of brushes and brooms, squeezing up against a boat filled with sacks of cement and another full of sand. The water level laps dangerously close to the gunwales of a passing boat overloaded with bricks. From further downstream drifts the sweet smell of thousands of pineapples, unloading from a similar gaggle of open 'country' boats.

A dealer might sell anywhere from 20 to a couple of hundred sacks of rice a day, usually by the 40kg (85lb) maund. A buyer turns up, inspects and sniffs the samples, dickers on the price and makes an offer. A middle quality rice might cost about US$8 a maund, but minighat, the 'rich man's rice', can cost up to US$12 to US$15. The deal is made, the invoice prepared, the buyer counts out and hands over the taka notes and continues on his way. His labourers will come by later to collect the purchase, usually in those standard 40kg sacks although sometimes rice is sold by the quintal, a 100kg measure.

The activity is just as intense out on the river as on the river banks. An endless procession of boats, from tiny one-man craft, no bigger than canoes, to large freighters, moves up and down the river. Open-decked passenger ferries, so overloaded that it's strictly standing room only, putt-putt up and down river. A steady stream of smaller ferries shuttle back and forth from ghats on opposite banks of the river, propelled by an oarsmen standing at the back and wielding a single paddle. A bright orange paddle-wheeled 'Rocket' ferry, the name *Mashud* posted across the side, arrives from some river port a day's travel away, its horn moaning sadly to announce its arrival. Blankets of water hyacinths float by downstream. From across the river there's the cacophonous bang and clatter of hammers and tools on metal as ships are refurbished and overhauled. Occasional flashes from welding torches add further confirmation of all the work going on.

There's more wholesale activity in the Mohammedpur Wholesale Rice Bazaar, just beyond the National Assembly building in the centre of Bangladesh's crowded capital. It's a utilitarian concrete building with alcoves down each side, open at each end. Each alcove in the shadowy building is occupied by a rice trader, patiently sitting behind a table on which the samples, small mounds or bowls of the 12 to 16 various grades of rice, are arrayed. Behind the trader and down each side of the alcove are stacked great piles of 85kg sacks of rice.

Trucks loaded down with hundreds of sacks deliver rice which, in turn, is carried away by grossly overloaded pickup trucks carting off 20 to 30 sacks at a time; equally overloaded *vans* and simple bicycle rickshaws. The warehouse wallahs load and unload the vehicles with lethal looking hooks, hulking the sacks onto antique looking balancing scales.

Labourers hoist a bag of rice on to a boat hauled in to the riverbank.

This busy market stall in Cantho, in Vietnam's thriving Mekong Delta region, has more than a dozen different types of rice for sale. The rice vendor checks one of those rice varieties, but her assistant clearly feels she has sold quite enough rice for one day.

VIETNAM

Markets in the Mekong Delta

Cantho on the Mekong Delta is in the heart of Vietnam's most export-oriented rice-growing area, but there's plenty of rice for local consumption as well. A stall in the local market might have a dozen or more different varieties of rice on offer with prices varying from as low as 4000 dong to as much as

16,000 dong for unusual ultra-high quality varieties. The rices shown here are priced between 30 and 50 cents a kg.

Only six km from Cantho the floating market at Cai Rang is the biggest in the delta. It's very picturesque, not at all a tourist show like the well known floating market in Bangkok. Long tail boats, with a propeller right out at the end of a long driveshaft, run from a gimbal-mounted car engine, and more sedate row boats weave in and out of the chaotic scene. Those long tails are dangerous! The current here is fast flowing and although most of the market boats are downstream from the bridge the rice boats are all just upstream. Generally, the people on the small boats sell to people on the big ones, although with rice the trade seems to be the opposite way. Perhaps people come to sell vegetables and fruit, then buy their own rice.

Everything imaginable that can be loaded onto a boat is for sale. One has woven baskets, another deals in scales, there's even a boat with soft drinks and another with food so that market traders can grab a meal before getting back to business. Bay Tinh, in her open boat, has been at the market for about four years; her husband is a fisherman. Her most popular rice varieties are priced around 20 cents per kg. Customers pull up beside her boat, look and feel the rice, then taste it before getting down to negotiations.

Bay Tinh sells rice from her open boat at the floating market in Cai Rang

THAILAND

Frantic Business in Suphanburi

A truck pulls off the highway and into the entrance to the large Suphanburi Agricultural Central Market, about 70km north-west of Bangkok. A half dozen men and women rush towards the truck wielding hollow circular-bladed knives. In what looks like a frenzy they stab the knives into the netted sides of the truck bay and rice grains spill down the hollow centre of the blade into the handle. Uncapping the handle they let the grains fall into their free hand.

Studying their handfuls of grain they rush to the front of the truck.

Rice buyers 'stab' an incoming truck to sample the rice. A handful of grains spill down the hollow knife blade into the palm of the buyer's hand, an instant assesment is made of the rice quality and a price shouted out in a quick bidding contest.

'*Saam phan rawy* (3100) baht', about US$75 – yells one of the men.

'*Saam phan sawng rawy* (3200) baht' – shouts one of the women.

They're bidding for the truck's contents and at 3400 baht the short and furious contest tops out. The successful bidder writes his price on an order form and hands it to the driver, who continues into the compound to have his truck weighed, drop its content into the successful bidder's open warehouse area and then have his now empty truck reweighed. The price bid and paid is per tonne of rice.

Mongkon Jongsuwan, the central market's rice expert, takes a handful of the rice from one of the buyers and glances at it disdainfully.

'Not very good,' he announces. 'The moisture content is too high, about 24 to 26%. And look how many of the grains are green. The rice usually isn't so good at this time of year, there's too much rain in the last month of growth. Next month the rice will be much better.'

With the next truck things are already looking better, the bidding starts at 3400 baht and the successful buyer closes the deal at 3600 baht. A few trucks later it's jackpot time. One glance at his handful of rice and a confident buyer offers 3900 baht; nobody tries to top his bid.

'This is good rice,' pronounces Mongkon. 'Long grains and the moisture content is low, I'd say about 14%.'

Rice samples are spread out on the table for quality checking with their moisture content written in chalk beside each sample. The price drops a couple of dollars a tonne for every 1% increase in moisture content.

Riceland International in Thailand

On the banks of the Pa Sak River in Ayuthaya, the ancient Thai capital about 80km north of Bangkok, Riceland International, mills and processes rice for the local market and for export to South and West Africa. The export rice is parboiled, a process which is not to Thai tastes. Rice comes in from central and lower northern Thailand and is shipped out by barge to be trans-shipped from the Bangkok port on the Chao Phraya River.

As each truck rolls into the Riceland plant a sample of the rice is examined for quality and checked for moisture content. Phuntipa Sriprasert was born and raised in this rice growing area and despite her youthful looks her understanding of rice quality was learnt from long experience, not textbooks. An experienced eye tells her the moisture content of a rice sample just as accurately as her testing equipment.

Phuntipa Sriprasert expertly assesses the quality of samples of rice while Kunyarat Ngamnimit gives a second opinion.

At the Karnal Mandi rice is displayed in great piles, waiting for rice dealers' inspections.

INDIA

A Punjab Grain Market

Punjab is the ricebowl of India, an area reputed for its high quality as much as its high production. In Punjab a grain market is a *mandi* and they're big, colourful, busy and crowded places with the rice dealers lined up around an open area where rice is heaped in great piles waiting to be bagged. At the height of the season the rice markets may operate 24 hours a day, but generally the farmers arrive with their rice early in the mornings and the dealers start their buying rounds from 2 pm. Peak buying activity is around 4 pm, as the dealers move from seller to seller making offers.

There's a floor price for unscented rice, guaranteed by the government, but not for the more expensive scented rice. Rice bought by the government goes into their buffer stocks, currently standing at around 45 million tonnes for all food grains. Dealers buy directly from the farmers but sales are also made to farmer co-ops which, in turn, buy fertiliser, seeds and other agricultural inputs. Much of the fine Basmati rice is sold directly to mills with which the farmers have established a relationship. Quality, fluffiness, non-stickiness, scent and sweetness are all factors which influence a price for the rice.

A Punjab Rice Dealer

Sitting outside his 'rice dealer and commission agency' in the Karnal Mandi, Parmod Kumar Gupta muses about his years in the rice business. 'How long have I been dealing in rice? Since the day I was born. My forefathers were rice dealers, I simply followed the family tradition.'

'This is not a good year,' he continues, 'expenses are costly and demand is low.'

'The quality is good,' he reluctantly admits, 'but the price is not high enough.'

'It's all for local use,' he goes on, 'there's no export demand, they all want Pakistani Basmati.'

Nevertheless he looks comfortably solid as he surveys a large heap of rice in front of him. 'Once the rice is bagged, trucks come and take it to my *godown* (warehouse),' Mr Gupta concludes, 'in a good year I buy and sell about two *lakh* (200,000) bags of paddy. Rice is my main business but I deal in wheat as well.'

Rice dealer Parmod Kumar Gupta.

Following page: Warehouse workers clamber over sacks of rice in a warehouse beside the Grand Trunk Road, the major route across north India since before Kipling's time. Rice mills and rice storage facilities dot the Grand Trunk Road all the way from New Delhi to the Sikh's holy city of Amritsar, close to the border with Pakistan.

JAPAN

Modern & Traditional Japanese Rice Shops

In Japan rice is usually bought in a supermarket these days but there are specialist rice shops like this one, simply signposted 'Rice Centre'. It's owned by the local farmers co-operative and sells rice from all over Japan – a map shows where the various varieties come from. The most expensive is the fine Koshihikari from the slopes of nearby Mt Tsukuba. The rice comes pre-packaged in colourful five, 10 and 30kg packs but is also available in bulk. Whichever way it's purchased it's likely to be some of the most expensive rice in the world. A kilogram of ordinary rice typically costs 30 cents or less in most Asian countries, perhaps 50 cents to as much as a dollar in the US or in Western Europe. In Japan it's typically 10 times as expensive as elsewhere in Asia and the fancier varieties can be twice that.

Yukio Nagatsuka has been a traditional rice dealer in Kome Ya for 30 years. He buys directly from the farmers, mills and bags the rice himself and sells directly to customers who want rice from a specific area. It's a dying trade as more and more of the rice business moves to supermarkets while connoisseurs who want a certain type of rice often buy directly from the farmers. His rice sells for around US$40 for a 10kg bag, a price pretty much in line with the supermarkets.

Kazue Onogi prepares tubs of bulk rice in her Rice Centre

Yukio Nagatsuka checks the paperwork in his traditional rice shop

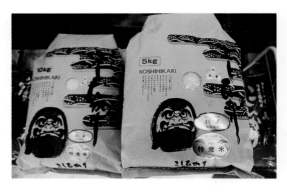

Fine Koshihikari – 'the best rice is almost transparent, whitish rice is not so good,' report the rice connoisseurs.

A box of rubber stamps to mark packages of rice intended as gifts.

Using

One rice eater's perfectly cooked and nicely sticky bowl of rice may seem to another overcooked and almost gluey. It's all a case of rice preferences. There are said to be over 100,000 different varieties of rice and there are certainly wide differences in the personal favourites of each rice eating country.

The Japanese, for example, like their rice to be a little sticky so that it's easier to pick up with chopsticks and, equally important, so their sushi doesn't fall apart. The lower the *amylose*, the starch part of the rice, the softer and stickier the rice will be. The Koreans and Chinese also prefer their rice sticky, with amylose of 10 to 18%. At the other extreme rice eaters from the sub-continent – India, Pakistan and Sri Lanka – like every grain to be distinct and separate. They're used to picking up rice with their fingers and their perfect rice will be 25 to 30% starch.

In South-East Asia, and for that matter for American and European rice eaters as well, the preference falls between the two extremes. In Laos and amongst the hill tribes of northern Thailand you'll find a taste for extremely sticky rice with an amylose figure down around 2%. No matter how sticky the rice, however, you should still be able to identify and separate every single grain. If it's merged together into one amorphous mass, that's not extra sticky rice, that's simply badly cooked rice.

How much rice does a rice eater consume? More and more as they get richer, then less and less as they get richer still. Once a nation becomes so wealthy that its citizens can consume all the rice they want, the demand immediately starts to fall. Increasing affluence simply leads to more diversified tastes and demand for other staples like bread, pasta or potatoes as well as foods which, in developing countries, are real luxuries, like meat and fish. So rice consumption has been declining in Taiwan, Japan and Korea. South Korea that is, in North Korea the curve would still be upward if demand could only be met. In Malaysia and Thailand the rice consumption curve has also topped out and is starting to drop.

Of course in developed countries where rice is not traditionally the main staple, rice consumption figures may reflect what is fashionable. In recent years rice consumption in the US, for example, has been steadily rising as the population develops an interest in Thai, Vietnamese, Chinese and other Asian cuisines. Today it's probably just over 10kg (20lbs) per person each year.

At My Hanh Bac, not far from Ho Chi Minh City, Nguyen Thi Tu turns out 800 rice paper sheets a day for that essential component of Vietnamese cuisine, rice paper rolls. First rice is ground to make a white liquid which is then grilled using the rice husks as the fuel. Finally the sheets are peeled off the grill and spread out on bamboo frames to dry.

VIETNAM

Rice Paper Rolls

Not all rice arrives on the table as a bowl of individual grains. It's often processed into a variety of final forms and nobody finds as many different ways to use it as the Vietnamese. *Bánh tráng*, rice paper rolls or rice paper wrappers, are an essential component in Vietnamese cooking, as familiar a dish as spring rolls in Chinese. They are made from rice flour which are laid out on bamboo frames to dry in the sun. The mesh of the frames gives the rolls their distinctive cross-hatched look.

Rice paper rolls are far from the only use of rice in Vietnamese cuisine. While the Chinese make their noodles, and their spring rolls for that matter, with wheat, the Vietnamese use rice. The thin vermicelli-like noodles are known as *bún* and are used in soups and noodle salads. The super-thin noodles, like angel hair pasta, are *bánh hoi*. Then there's *bánh pho*, rice sticks which are stirfried or added to soup.

Rice powder can be roasted to produce *thính* which adds a slightly bitter taste to meat dishes or rice vinegar, *dám gao* can be used in marinades. You can even drink it as *ruou can*, although that's a surprisingly familiar use of rice. The Balinese are also well known for their rice wine but rice even finds its way into beer. Or at least into Budweiser, German regulations insist that it's only beer if barley is used, so in Germany Budweiser is not beer.

The rice paper rolls are spread out on bamboo frames to dry.

Indonesia — Rice cakes, wafers and offerings in Bali.

Vietnam — At a Mekong Delta ferry crossing in Vietnam a stall sells sweet rice snacks neatly
packaged in a banana leaf to waiting motorists. Any snack like this, wrapped up in a leaf, is known in
Vietnam as banh, a word which translates, not very satisfactorily, into English as 'cake.'

JAPAN

Tatami Mats

Yoshio Otani runs the last traditional tatami factory in Ibaraki-ken, the prefecture just north-east of Tokyo.

'Perhaps,' he muses, 'Otani Tatami is the last traditional mat maker on the whole Kanto Plain.'

Tatami mats are the closely woven floor matting used in Japanese homes, the sacrosanct area where shoes are never worn. Room sizes are traditionally expressed in *jo*, the size of a tatami mat, although today there are older, larger *jo* sizes and newer, smaller ones. A tatami mat is a three dimensional panel rather than just a two dimensional floor covering and the traditional version is 6cm deep. The reed outer layer of a tatami mat, known as *goza*, with its cloth edging is just the tip of the iceberg, beneath lies a deep cushion of tightly packed rice straw.

With Superman's X-ray vision you'd see a very different story beneath a typical modern tatami mat. Compressed rice straw has given way to a high-tech sandwich of styrofoam and other modern materials. At the same time tatami mats have been getting thinner, modern versions may only be a couple of centimetres thick. There are a variety of reasons for this manufacturing shift. For a start it's much easier to make a tatami mat with modern materials rather than rice straw. They don't have to be carefully handled and dried, they don't have to be treated against insects and they're much neater

and tidier to work with. Furthermore, the end result is much simpler to transport and install, a traditional tatami mat weighs in at a hefty 35kg, while its modern competitor tips the scales at just 5kg. As a final disadvantage for the traditional rice straw mat, the rice straw itself is no longer easily available. Modern combines chop the straw up too small to be used. As a result only 10 to 15% of new tatami mats are made with traditional rice straw. The rest have the modern synthetic core.

The vast majority, perhaps 80 or 90%, of traditional tatami mats are no longer made in Japan – they are now sourced from North Korea. So Yoshio Otani's small enterprise, it employs just 18 people, including his two sons, is the tiny remnant of a once proud industry. Apart from making their own tatami mats, Otani Tatami also imports mats from North Korea, where they have helped install Japanese tatami mat manufacturing equipment. It has become so hard to find suitable rice straw in Japan that even their Japanese-manufactured mats are principally made from imported North Korean straw. Much of their manufacturing is of special made-to-order mats carefully sized to fit specific floor shapes. It can take a whole day to turn out a single one of these customised mats.

Still, Yoshio Otani reflects, there's continued demand for the real thing. He tells of building workers pretending to stagger in under the weight of a genuine rice-straw tatami mat, when actually all they're carrying is a modern substitute.

Packing down armfuls of rice straw to make a tatami mat.

Yoshio Otani works on a made-to-order traditional tatami mat, authentically manufactured from rice straw.

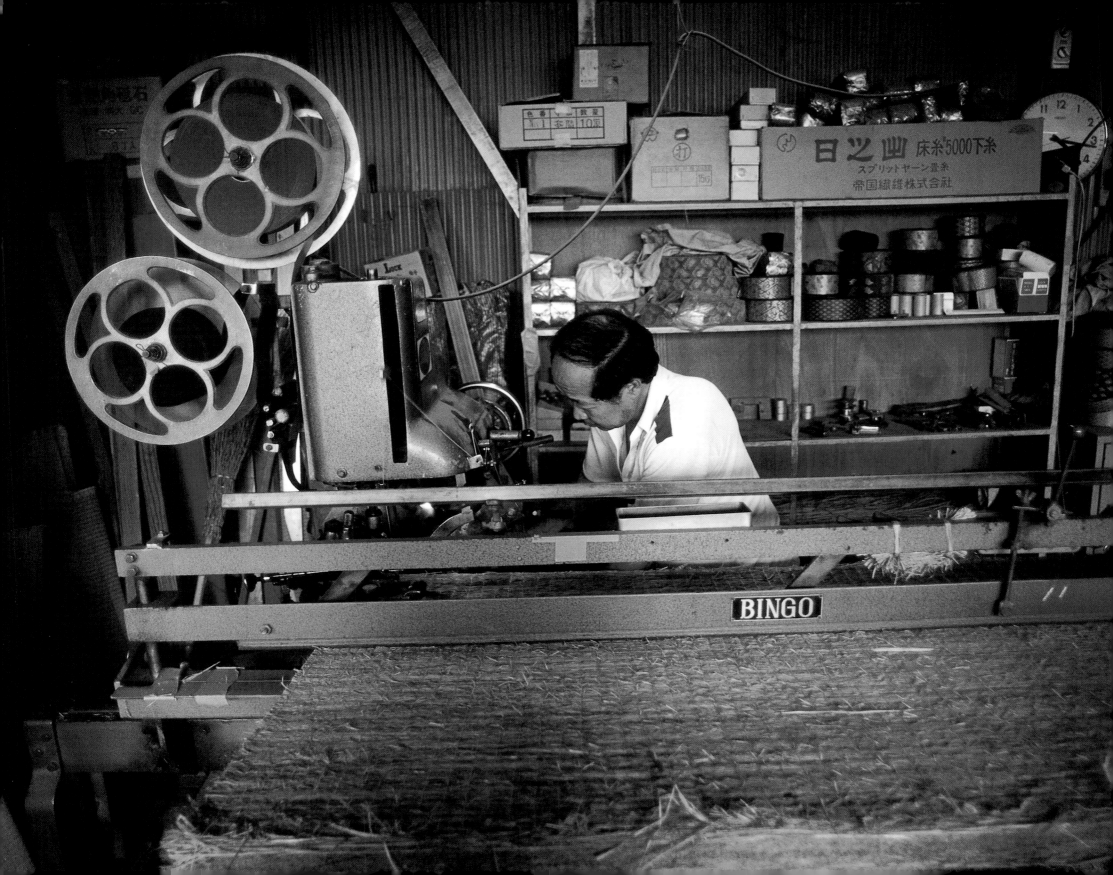

Rice Giving, Rice Eating

In Buddhist countries like Thailand and Burma monks start each day by collecting alms, walking quietly through their town or village, often in single file. Usually their bowls will be filled with rice. Back at the monastery the food is shared out and eaten communally, it may be their only meal of the day. The practice follows that of the Buddha and is thought to gain merit both for the rice-receiver and the rice-giver. Thailand has over 400,000 monks, every young Thai male is expected to spend at least a short period as a monk and each day a quarter of a million monks fan out to collect their alms.

Buddhism is equally important in Burma, the world leaders when it comes to rice consumption. The average Burmese chows their way through about 195kg (430lbs) of uncooked rice every year. People in other countries in Asia where rice is the main staple also eat a lot of rice, although not quite up to Burmese levels. In Bangladesh, Vietnam, Indonesia or Thailand average consumption is around 130 to 150kg (290 to 330lbs) a year. India falls well behind, perhaps because the north Indians like their delicious breads as much as the south Indians like rice.

South Korea and Japan were probably right up there with those big rice eaters a few decades ago, but not anymore. Today the average South Korean eats about 100kg (220lbs) of rice a year, the average Japanese less than 60kg (130lbs). This is a pattern that's likely to repeat across Asia. Increasing prosperity brings with it increasing demand for rice, but only up to a certain level.

Burma – At Kyaiktiyo, home of the famous balancing pagoda, Buddhist monks on their morning round line up for donations of rice.

Clockwise from top left

Burma – Chicken briyani at the Nila Briyani Restaurant in central Rangoon.

Bangladesh – Stirring a large bowl of rice in a Dhaka restaurant.

Singapore – Abu Bakar stirs 25kg of fine Pakistani basmati rice in the Victory Restaurant.

Indonesia – White and yellow rice Denpasar market, Bali.

Singapore – Hainanese chicken rice, a favourite dish in the city-state.

Celebrating

In Asia each stage of the rice cycle from planting to harvest is a cause for celebration and nowhere with more frequency and style than in Bali. Every Balinese village has at least three pura or temples: for the village founders, for the spirits of everyday life and for the spirits of the dead. At the very least a temple has to be honoured with a 'birthday' festival or *odalan* every Balinese year and since the year is only 210 days long that means an awful lot of festivals. Offerings of fruit, flowers and food, which always include rice, are a highlight of these temple festivals.

Furthermore, every Balinese family compound has a small – or sometimes not so small – temple for the family, and since rice is of such great importance and the distribution of irrigation water to the rice fields is handled with such care there are also rice growing temples, known as *pura ulun suwi* or *pura subak*. As we have seen, rice growing in Bali centres on the *subak*, the farmer's irrigation association. There will be a *subak* temple at the spring that is the source for an association's irrigation network, and smaller temples and shrines at every point along the chain, all of them requiring a continual cycle of rituals, offerings and ceremonies. Particular respect is paid to Dewi Sri, the Goddess of Rice.

The rice growing cycle is treated with equal respect in Thailand, starting each year with a ploughing ceremony which helps to predict the size of the crop. Then there's a ceremony to ensure the all-important rains arrive on schedule and offerings will be made to local monks from the first harvest. Intermingled with the larger ceremonies will be small offerings to the spirits of the rice fields and in particular to Mae Phosop, the Mother Spirit of Rice, and Phi Ta Haek, the guardian of rice field fertility. In India and Nepal, Lakshmi, the Goddess of Wealth and Prosperity, is also honoured with rice.

We encountered other rice growing ceremonies as we followed the rice trail. In Cambodia the royal ploughing ceremony was watched over by King Sihanouk himself. In Japan, rice's all pervading importance was underlined by the salarymen, Japanese office workers, joining in a harvest festival we chanced upon. In the Philippines a festival honouring St Isidore, the patron saint of farmers, included beauty queens as part of the festival procession. Even Australia gets into the mood every second year at Leeton's Easter rice festival.

Philippines – The Pahiyas Festival at Lucban, south of Manila, features these colourful kiping, edible decorations made from brightly coloured rice flour wafers 'fried' in kabal leaf moulds.

Harvest festival at temple in Pejeng near Ubud, Bali.

INDONESIA

Offerings

Religion permeates every aspect of Balinese life, from the temples found in every village to the small shrines which dot rice terraces. The island is full of spirits, good and bad, and to keep the good spirits on side and the bad ones at bay requires a non-stop cycle of offerings and ceremonies.

Step out of your hotel bungalow in the morning and there, outside your door or at the junction of the path, is a pretty little offering lying on the ground, perhaps woven of palm leaves or cut from a larger banana leaf. The fact that it's lying on the ground is a clear indication that this was not intended for a good spirit, it's down there for the spirits who dwell at ground level, the demons. In fact it's a bribe, a pay-off. Put an offering at your door and with luck the demon will take it and not come inside. Nobody wants demons in the house.

Look closer at the offering and you'll find that its position is not the only thing which identifies it as something intended for a bad spirit. Demons and evil doers like strong flavours and so their offerings may contain a drop of arrack, some grains of salt, even a chilli pepper.

Now look up. On the window sill, in an alcove, on a platform by your door or even in a special shrine, there will be more offerings. Once again position indicates what sort of spirits they are intended for – good spirits don't hang around at foot level. The contents of these offerings will also indicate their end users. Good spirits like pretty things, nice smells, sweet flavours – so an offering for a good spirit might include flower petals, some sugar, a slice of banana.

Far from dying out as Bali becomes more affluent, educated and modern, the practise of putting out offerings seems to have accelerated and intensified. If you're doing well there's more to be thankful for. And more to lose, which makes keeping the nasty demons at bay even more vital. The modern world also brings greater dangers and more possibilities for things to go wrong. Wander down the narrow alleys, known as *gangs*, which criss-cross the Balinese beach resort of Kuta and at every junction you'll see piles of those demon offerings. Motorcycles accidents don't happen by chance, they're just as likely to be caused by unbribed demons as by riding too fast.

Good or bad the offerings will include rice. Whatever its intention a spirit has to eat after all; but not too much rice for the demons, you don't want them getting too strong.

Rice is thrown on to the cremation pyre at a funeral in the village of Tegalalang, Bali.

Prince Norodom Sirivudh, half brother of King Sihanouk, is carried by palanquin for the short distance from the Royal Palace to the ploughing ceremony.

CAMBODIA

Royal Ploughing Ceremony

Presided over by King Sihanouk and other members of the Cambodian royal family, *Chat Preah Nengkal* celebrates the ploughing of rice fields and the planting of the new rice crop. The colourful festivities take place in the field in front of the National Museum and beside the Royal Palace in Phnom Penh.

Down one side of the field are contingents of men in bright orange outfits along with groups from the Cambodian army (khaki), navy (white) and air force (blue). Although one prince

is carried in by palanquin, other dignitaries arrive by car, accompanied by police motorcycle outriders, flashing lights, sirens and brigades of marching school children. King Sihanouk also eschews the traditional man-powered transport and arrives in a stretch Mercedes Benz, with Queen Monineath waving regally.

After a lengthy musical interlude, chanting and speeches, three pairs of oxen set out for four circuits around the ploughing ground, led by the prince and followed by his princess in the palanquin. The oxen are beautifully turned out, their regalia topped by red horn covers. They may be the most elegant oxen ever to pull a plough but the ploughmen,

A symbolic ploughing of the field by the most ornately decorated oxen ever to get near a rice paddy.

After the symbolic ploughing Queen Monineath, sheltered by a golden parasol, does some symbolic sowing of the rice seed.

The ceremonies concluded there's a mad scramble to grab the remaining grains left by the oxen.

resplendent in white jackets and purple jodhpurs, are a fine match. In fact the ploughing is purely symbolic, the plough never touches the soil.

The oxen are followed by small groups of women in gold, red and green and then a large entourage of women wearing orange pantaloons and white tops. With flags, banners, gold poles and parasol-toting retainers to add another dash of colour it's quite a parade.

The ploughing concluded a small ceremony takes place, overseen by a worried looking gentleman, sweltering in the tropic heat in a dark suit and tie and clutching his mobile phone. Finally two of the oxen are led over to seven vessels

containing a variety of grains, seeds and liquids – predictions for the coming rice-growing season can be gleaned by watching which container the oxen prefer.

Next day the Reuters report on the oxen-led prophesy announces that they ate the rice, beans and maize, which should ensure the best harvest for two decades. Furthermore, the abstemious animals steered clear of the vessel containing rice wine, an indication that, 'we'll have peace, no violence,' said palace priest Din Proum. There's loud applause as King Sihanouk leaves the gathering, sheltered by a huge orange parasol.

Hordes of sweaty salarymen race around carrying impressively large portable shrines known as mikoshi and shouting 'rashoi, rashoi,' (let's go, let's go) at the Tono Harvest Festival.

Women at the Tono Harvest Festival parade.

JAPAN

Japanese Rice Festival

The Japanese are very enthusiastic about their festivals, known as *matsuri*, which feature colourful and noisy parades. In September there are often festivals celebrating the rice harvest, such as the Tono Harvest Festival in northern Honshu. The city of Tono was formed by merging eight villages in the 1950s. It's in a region where traditional farmhouses, *magariya*, are still found, where farming families and their livestock live in the same building.

Carved image of a bulul, the primitive-looking rice god figure set up to guard a rice granary in the Ifugao ricelands of North Luzon.

PHILIPPINES

Rice Gods

Bululs are primitive looking figures, male and female, usually carved in a squatting position with hands folded on knees. A *bulul*, or more correctly two since they should be male-female paired, should guard every rice granary. Unfortunately many old *bululs* are now in museums or have been bought by overseas collectors so many *bululs* are now solitary. It's estimated that there are only about 300 original *bululs* left in Ifugao territory. If a family has more than one granary they can have more than one *bulul*, or pairs of *bululs*. Although possession of a rice granary is supposed to be the basic requirement for having a guardian rice god in fact anybody who has sufficient area of rice fields to fill up the rice attic in their house is deemed worthy of owning a *bulul*.

These rice god figures often have mythical abilities, they may be able to prevent rats from getting into the granary or even be able to kill them if they have gained entry. There are stories of *bululs* mysteriously escaping from houses that catch fire. More important bululs are capable of increasing the quantity of rice harvested or even the quantity of rice stored in their granary. Usually *bululs* are only brought out at harvest time when they are positioned under the granary and guard the rice when it's left out to dry, 40 days being the traditional drying period. When the rice is stacked away in the rice granary or in the rice attic in the house the *bululs* are put away with it. A family's *bululs* are inherited by the first child, who also inherits the rice fields.

The village of Bucos, right across the river from Banaue has several *bulul* carvers as well as a famous old *bulul* only brought out once a year for a harvest festival.

A bulul image guards a rice granary door.

Partially carved wooden bulul statues in the village of Bucos, just across from the river from Banaue.

Following Page: Ranks of bululs are carved on a ceremonial chair panel.

PHILIPPINES

The Pahiyas Festival

South of Manila the Pahiyas Festival in the small town of Lucban honours San Isidro (St Isidore), the patron saint of farmers. House facades throughout the town are decorated with agricultural produce for this major thanksgiving festival in mid-May. Pahiyas means 'deck the halls' and there's intense competition to have the best decorated house. A procession winds its way through the town in the afternoon; the route varies from year to year so that everybody in town can join in.

The procession features everything from bands decked out in red and white outfits and pumping out Rock Around the Clock to the inevitable beauty queens including Miss Rice and Miss Sunflower. There are also lots of farm animals in the parade, introducing some inevitable difficulties. Beauty queens, trailing long dresses, do not usually have to worry about putting a dainty high heel into reminders that a carabao, the native water buffalo widely used for ploughing fields in the Philippines, has recently passed by. As well as the many decorations made with *kiping*, edible decorations made of brightly coloured rice flour, rice also plays a part in many of the parade costumes.

On the morning of the festival two men work at decorating their house.

At the Pahiyas Festival in Lucban, south of Manila, many of the harvest parade costumes feature rice, like this young lady's rice stalk hat.

Religion and rice, a group of Biblical
figures stand with their rice harvest and
a background of bright pink kiping.

There's usually something interesting cooked up as part of the festival, on this occasion
it was Australia's largest fried rice.

AUSTRALIA

A Rice Festivel in Leeton

On even numbered years the SunRice Festival features parades, performances, competitions and, with Australian style, an attempt to cook something very big indeed. The 10 days of festivities culminates with the parade on Easter Saturday and festival finale on Easter Monday.

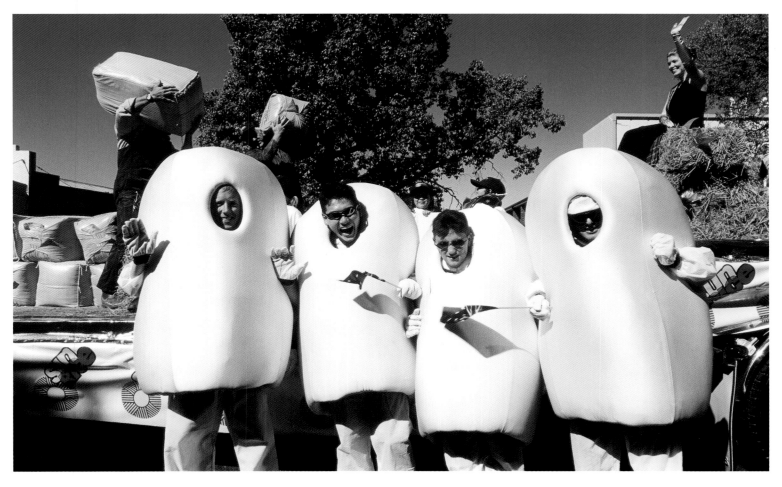

Happy rice grains join the SunRice Festival parade in Leeton, New South Wales.

Researching

The 'Green Revolution,' the agricultural upheaval which steered the world away from what looked like a Malthusian disaster, centred around the two most important crops - wheat and rice. It kicked off in the early '60s and not only achieved stunning results but rolled them out with amazing speed. In the late '50s it was widely feared that Asia's booming population was set to overtake and then race away from the region's ability to produce its most important staple food - rice. Responding to this disastrous scenario the Ford and Rockefeller Foundations funded the establishment of IRRI, the International Rice Research Institute.

The institute started work in 1962 and within a few years produced IR8, the first of many 'new varieties.' Known as semi-dwarf rice, the new plants were more productive (so farmers could produce more rice from the same size field), shorter and sturdier (so they could hold up their heavier load of grains), grew faster (so farmers could produce more crops per year) and even managed to be more disease resistant.

Farmers quickly adopted the new varieties and production was soon increasing more rapidly than population. In 25 years the world's rice production doubled and in some areas, like Indonesia, actually tripled. Today new varieties constitute 75% of the rice grown in Asia and although Latin America and Africa lag behind, in the USA and Australia the rice is 100% new varieties. IRRI's success led to the formation of CGIAR, the Consultative Group on International Agricultural Research, which coordinates research across a wide field of agricultural, forestry and fishery organisations.

This farming miracle was achieved at remarkably low cost. Even at its funding peak in the late '80s and early '90s, IRRI's annual budget rarely exceeded US$40 million. Today, annual funding has dropped well below US$30 million and, like many research institutions, IRRI is in a financial crisis due to a 40% global decline in support for agricultural research. Worldwide more than 1000 times as much is spent on pet food as on agricultural research!

Despite this success the world's population has kept on growing and once again threatens to overtake production. The search is on for rice varieties that have not just more tonnes per hectare but also a higher 'grain to straw' ratio. Currently a rice plant is about 50% grain, 50% straw; improving that split to 60:40 is one of the targets.

Thailand – At the Prachinburi Rice Research Center three workers – Ung, Mai and Aw – check an experimental plot of deepwater rice.

Reducing the amount of herbicide (weed killer) and pesticide (insect pest killer) needed is another target. Since wild rice species have managed to survive without farmers lending them a helping hand, scientists are hoping to introduce some of their natural immunity to farmed rice. A key reason for the labour intensive transplanting of rice in Asia is because transplanted rice has a handy headstart over encroaching weeds. The farm-to-city exodus in many regions of the developing world often leads to severe rice field labour shortages and new varieties that are more competitive against weeds will mean rice can be sown directly.

Improvements are not limited to the plant. Planting seedlings more accurately will make it possible to squeeze in more plants per hectare. Faster harvesting will mean more rice can be brought in at the ideal time while improved equipment can reduce grain losses. Accurately levelling fields not only minimises the amount of water used but also ensures that the whole field is flooded to the optimum depth thus increasing the yield. Even simple reliability can help, when farmers are uncertain that irrigation water will always be available they are prone to flood their fields to more than the optimum depth when it does flow.

Since water is such an important component in the rice story it's clear that there are huge possibilities for water usage improvements. Increasing demands for water - for domestic and industrial use as well as for agriculture - will make water scarcer and more expensive in the future. Improving the efficiency of water use will not only raise a farmer's profitability but may also make it possible to grow more rice, since production is often limited by the availability of water. A new IRRI strategy is to develop an 'aerobic' system in which rice plants grow the same way as irrigated upland crops such as wheat and corn. The aim is to realise yields of up to six tonnes per hectare using only half the water required in current lowland practices.

Better rice plants, improved techniques and superior equipment are all part of the story but at the end of the day the rice farmer is the key to the puzzle. Fortunately IRRI and other rice researchers are uniform in their praise for farmers' adaptability and interest in adopting new techniques and better plants. It's easy to see that pursuit of efficiencies when a farmer pores over computer yield readouts and ponders the value of using a GPS (global positioning system) to control fertiliser use, but developing world farmers are equally ready to adopt improvements. The quick spread of IRRI's new varieties is a prime indicator of their close attention to productivity.

PHILIPPINES

Cross Breeding at Los Baños

In an experimental greenhouse at the IRRI Los Baños centre, rice varieties are cross bred in the ongoing quest to develop strains with higher yield (more grains per panicle) and improved disease resistance. Much experimentation is being done on crossing *Oryza sativa* varieties with wild rice. Wild varieties are a vast and valuable genetic resource, which can provide genes for pest and disease resistance as well as other important characteristics. By crossing wild rice varieties with cultivated rice it is hoped to develop new strains combining high yield with greater resilience.

After checking that it has not yet pollinated, the top end of individual grains of rice are snipped off, emasculating the grain by removing the six anthers, the male portion, from each grain on the panicle. This must be done between 3 and 6 pm, before pollination takes place in the evening. The grains can be pollinated 24 hours later by waving a panicle of the desired cross breed over it, usually in a warm room under 200 watt lamps.

Noel Llanza and Nestor Ramos check the sample cross breeds, typically 100 to 150 experimental crosses will be made each season.

Left:
Alvaro Pamplona, an assistant scientist, works on cross breeding.
Middle:
A vacuum is used to suck the pollen out of the grain.
Right:
The panicle is covered with a small envelope and dated.

The IRRI Gene Bank

The IRRI centre houses the world's largest single crop gene bank with 108,000 varieties of rice in active storage at 2 to 4°C and base storage at -18 to -20°C. Active storage consists of samples in large foil packs plus two smaller foil packs which can be sent out instantly when seed requests are made. The base storage is in a large freezer room within the active storage area. Here the seeds are kept in sealed cans that should be good for 50 to 100 years. A second back up storage is kept at Fort Collins in Colorado, USA.

The vast majority of the varieties, about 102,000 of them, are the familiar Oryza sativa but there are also about 1600 different O. glaberrima varieties, the type of rice found predominantly in Africa, and about 4500 samples of wild rice. Except for a handful of samples of related genera, all the wild rice is of the Oryza genus, not the Zizania genus known in North America as wild rice. India has provided the single largest number of O. sativa types (15,000 of them) and of wild rice types (nearly 1000). Curiously India has also received the largest number of traditional native Indian varieties, repatriated to promote conservation in their country of origin.

The value of the gene bank has been amply proven in Cambodia. During the madness of the Pol Pot years not only were city dwellers forcibly relocated to the country but country folk were also shifted from one region to another and pushed into large collective farms. Often forced to abandon their homes and goods, farmers would always take with them the one absolutely vital contributor to their livelihood: their rice seeds. Cambodia's rice types became a hopeless jumble. Varieties carefully bred over the centuries to perform in the hills found themselves on the lowland plains. Rice which worked well in salty coastal air was grown far inland. Rice which coped with low rainfall was grown in the wettest areas of the country.

As a result when peace finally returned yields were still far lower than before Pol Pot. Fortunately IRRI had a complete stock of Cambodian rice types, each catalogued according to location. These rice varieties were reintroduced, matching the ideal variety with the appropriate location. The introduction of new varieties through IRRI's Cambodian outpost near Phnom Penh has brought the country back to self sufficiency in rice production.

Selecting and storing rice varieties in the gene bank is a labour intensive and exacting operation. First good seeds are selected from the samples provided by 15 to 20 mainly female seed sorters. There are currently 15,000 rice types in the process of being catalogued and stored. The selected grains are dried at 15°C and 15% relative humidity and a sample of seeds are tested. At least 85% of them must germinate or they must be recycled to create a new higher quality batch of seed. Sometimes only very small samples of seeds are received and in that case it's necessary to grow rice plants from the first sample in order to increase the number of seeds available.

Boy Almazan, an assistant in the IRRI gene bank, checks some of the 108,000 rice seed samples in the Los Baños base storage freezer room. In the background is Flora de Guzman, the senior associate scientist in charge of gene bank operation.

INDIA

Punjab Progress

'Purchasing power is still a problem in India,' reflected Dr T S Bharaj. 'We produce enough food for everybody but there are people who simply cannot afford it.' 'Pockets of poverty.' agreed Dr Malik Singh, the head of the seed production unit at the experimental seed farm in Narayangarh.

'But not here in Punjab,' added Dr Bharaj. 'Nobody goes hungry in this state,' he continued, patting his stomach, 'the Punjabis are all solid people.'

They are. By every measure Punjab is the best fed and most affluent state in India and it's not from luck or any natural advantage. In fact, apart from good farming conditions, Punjab has no natural advantages: there's no mineral wealth and the state probably suffered more damage than any other part of India in the violence that accompanied the partition of the sub-continent in 1948. The break actually split the Punjab region in two, half of it ending up in the new country of Pakistan. The state has reached its enviable position by a simple method – hard work. The Sikhs, who make up a large part of the state's population are famed for their no nonsense approach to life and their 'can do' attitude.

'Farmers are always approaching us about new seed varieties,' confirmed Dr Bharaj. That's a situation in clear contrast to many other rice growing areas around the world where scientists have to put as much effort into convincing farmers to try new varieties as they do in developing them.

'Punjabi farmers are also willing to invest if they can see a return from it,' went on Dr Bharaj. That's clear too, there's always a tractor in view as you drive along the back roads of the state, and Prakash Motors, the New Holland Ford tractor dealer at the Khanna grain market, has shiny new, blue painted tractors ready and waiting for delivery. Where much of developing Asia still harvests its rice by hand the Punjabis bring in 95% of their crop with modern combine harvesters.

Dr T S Bharaj and Dr Malik Singh inspect PR114, a very successful Punjab new rice variety

'I think one of the reason Punjabis are so willing to try new technology is that so many Sikhs have emigrated,' Dr Bharaj continued. 'Nearly everybody has a relation in America or England, when they come back they bring new ideas with them.'

We continued to the Central Soil Salinity Research Centre in Karnal where Dr B Mishra and Dr R K Singh were working on a new export quality Basmati rice which combined moderate tolerance to salinity with higher yield. India has thousands of square kilometres of farmland with salinity, alkalinity or coastal salinity problems and work continues on developing rice varieties that can cope with these difficult conditions. Irrigation is often a prime cause of these problems and better drainage and other improvements to the soil can help, but these changes tend to be expensive and farmers often don't have the cash. The centre has developed a number of new varieties with high tolerance for problem soils.

Top:
Dr D S Dodman and Dr S D Dhiman look at a handful of Haryana high yield unscented rice.
Middle:
Dr Rakesh Sharma and Dr Khushi Ram talk with Zile Singh, his ox cart laden with rice straw.
Bottom:
Dr B Mishra and Dr R K Singh in the greenhouse at the Saline soil research centre in Karnal.

Philippines – *From the Punjab to the Philippines the number of varieties of rice is simply amazing, this is just a small selection of the 108,000 different types in the IRRI gene bank.*

IR RICE TYPES

Outside the IRRI headquarters at Los Baños small plots of a number of historic IR rice types are grown each year, including the grand-daddy of them all, IR8.

By judicious application of fertilisers and careful cultivation farmers had been able to increase their yield only to see their crop, quite literally, fall over. When the panicles, the top of the plant where the grains grow, became too heavy the stalk simply could not carry the weight, it would bend over or 'lodge.'

The scientists first task was to produce a new variety which combined high yield with resistance to lodging and also grew faster so that farmers could produce more crops each year. They did this by crossing the tall, insect-resistant, tropical, Indonesian Peta rice with the shorter, high-yielding, Taiwanese dee-geo-woo-gen variety. The result was IR8, the first of the new 'semi-dwarf' rice types with short, stiff stalks which could carry much heavier panicles without toppling over. Not only was IR8 more productive it also grew faster, maturing in 130 days against the 160 to 170 of its predecessors. Released in 1966, it was instantly successful, effectively doubling the yield potential for Asian rice growers.

Ten years later the numbers had rolled around to IR36, which had a better grain shape, faster growth, was more resistant to disease and insects and even had improved cooking characteristics. Hardly surprisingly IR36 became the most popular rice variety in the 1980s, yet in 1985 it was superseded by the even better IR64. More than half of the rice area in the world is now planted with IR varieties or their progenies, but the IR series was discontinued in the 1990s as the emphasis shifted to national and local agricultural research centres.

In addition to developing new higher-yielding rice plant types that carry three or four panicles per plant and no unproductive tillers, IRRI research has also been breeding improved varieties with natural resistance to pests and diseases. This reduces the need for farmers to spray their crops with pesticides. Also being tested are a range of options in integrated pest management that further reduce farmers' need to spray potentially dangerous chemicals. Work continues on simple but reliable techniques for farmers to optimize their fertiliser use. Ensuring a crop's maximum uptake of fertiliser, by applying just enough of it at just the right time, means less fertiliser runoff polluting rivers and streams. The added benefit - and the immediate incentive for farmers - is lower input costs which in turn means improved income.

IRRI is also advancing the new concept of biofortification, or breeding rice varieties that have a higher nutrient content in the endosperm of the grain, the part left after milling. Iron, zinc and vitamin A are three essential micronutrients which are receiving the most attention.

Philippines – Outside IRRI's Los Baños headquarters a small plot is grown each season of IR8, the original rice type which sparked the 'Green Revolution'.

On the Rice Trail

Indonesia –Tony gets involved in transplanting Iseh, Bali, .

Indonesia – Richard photographing a recently transplanted field, Ceking, Bali.

India –Tony checks the figures with scientists from the Kaul Rice Research Station in Haryana.

Indonesia – Richard photographs the rice threshing operation at Pejeng in Bali.

Thailand – Tony inspects rice varieties at Riceland International in Ayuthaya.

Bangladesh – Richard supervises the rice loading operation beside the Buriganga River in Dhaka.

To trace the story of rice we visited, talked with, interviewed, watched and photographed rice farmers, rice processors, rice dealers, rice merchants, rice researchers and even, at the very end of the chain, rice consumers. We tracked these varied people down in 13 different countries, each one with distinct differences and points of interest. Our travels involved six separate trips and more than 30 flights on a dozen different airlines.

Remarkably the processes and patterns were often surprisingly similar despite the difference in scale and technology. The whole village of Ababi in Bali may have turned out and spent the morning hours bringing in a tonne of rice while Nick and Kerry Lowing in Australia used their huge and highly computerised 'header' to harvest 200 tonnes a day, but each rice farmer was equally concerned that their rice had just the right moisture content.

Turning unhusked rice into the final polished product involved exactly the same steps whether it went through a high tech rice mill in Thailand or a totally manual process in the highlands of the northern Philippines.

Shoppers were just as fussy about the rice they bought in a market in rural Burma as in a pricey, air-conditioned supermarket in Japan.

INDEX

GLOSSARY

Amylose – The starch content of rice.

Arborio – Italian rice, popular for making risotto, the best known Italian rice dish.

Aromatic Rice – Rice with a fine scent, usually basmati or jasmine.

Awn – spiky point at end of a rice grain's outer covering or husk.

Basmati – The fragrant long grain rice grown in India's Punjab state and across the border in the Punjab region of Pakistan. Held by many connoisseurs to be the finest rice of all.

Bay – In Australia an individual rice field is known as a bay.

Black Rice – Dark coloured rice found in some parts of Asia and often used in desserts.

Bran – Coating on the rice grain which is milled off, converting brown rice to white rice. The bran constitutes about 5% of the weight of unmilled rice and also makes up a large part of the nutritional value of rice. Rice bran is used as a food additive or for animal livestock feed.

Briyani – Indian flavoured rice dish combining north and south Indian influences.

Broadcast – Sowing rice by simply scattering the seeds in the field, rather than the labour intensive method of transplanting.

Broken Rice – The milling process inevitably breaks some individual rice grains, particularly if the moisture content is too low. This broken rice is still edible but is usually much less valuable than unbroken rice. High quality rice will have a very low percentage content of 'brokens.'

Brown Rice – Rice which has been husked but not milled. The brown colour comes from the bran coating and retention of the bran means the rice is more nutritious than milled or white rice. Brown rice takes longer to cook than white rice and is chewier.

Cargo Rice – Brown Rice, the state in which rice is usually shipped.

Chelo – Classic Iranian rice dish.

CGIAR – Consultative Group on International Agricultural Research, an association of public and private members supporting a system of 16 international agricultural research centres that work in more than 100 countries.

Coleoptile – See Radicle.

Combine Harvester – Machine that simultaneously cuts and threshes a crop of standing grain.

Converted Rice – US term for parboiled rice.

Deep Water Rice – Rice which can grow in water a metre or more deep.

Dehra Dun – Hill town in the Indian state of Uttar Pradesh reputed to grow the finest Basmati rice.

Dwarf Rice – See Semidwarf Rice.

Endosperm – The starchy part of the rice grain which is consumed.

Enriched Rice – Rice with vitamins or proteins added, ostensibly to replace those lost when the bran is removed in the milling process. In fact in a balanced diet the food value lost with the bran is unimportant but enriched rice is popular in the USA.

Floating Rice – See Deep Water Rice.

Flowering – Each panicle divides into multiple spikelets each ending with a flower bud. It is the pollination of the flower which leads to the rice grain developing.

Fortified Rice – See Enriched Rice.

Fully Milled Rice – See White Rice.

Glutinous Rice – The extremely sticky rice favoured by the hill tribes people of northern Thailand and Laos.

Header – The cutting and harvesting apparatus attached to the front of a combine harvester. The whole machine is often referred to as a 'header.'

Head Rice – Unbroken rice grains after the milling process. Milled white rice is made up of head rice and brokens.

Hull – See Husk.

Husk – Outer covering of the rice seed, milled off to convert 'rough' or 'paddy' rice to 'brown' rice which has to be further milled to produce white rice. The husk typically makes up about 20% of the total grain weight. Rice yield figures should indicate whether the rice is rough or husked. In the third world the rice husks are often burnt to produce steam for parboiling or to power steam-engined milling machines. The husk ash is high in silica and is used in industrial production.

Indica – One of the two main forms of *Oryza sativa* this is the main type of rice grow in the tropics and subtropics including Philippines, India, Pakistan, Sri Lanka, Indonesia, central and southern China as well as in African countries. Indica plants tiller profusely and have short to long, slender, somewhat flat and almost awnless grains. They shatter easily and have a 23 to 31% amylose content. See also japonica.

IRRI – The International Rice Research Institute at Los Baños, south of Manila in the Philippines, is an autonomous, non-profit agricultural research and training organization with offices in 11 other Asian nations. The Institute's main goal is to find sustainable ways to improve the well-being of present and future generations of poor rice farmers and consumers while at the same time protecting the environment. It was IRRI's development of IR8, the high yielding dwarf rice developed in the 1960s which was a prime factor in the success of the 'Green Revolution'.

Irrigated – The largest proportion of the world's rice crop is grown in irrigated fields where the water depth can be maintained at an optimum level. Irrigated fields comprise about 50% of the total area devoted to rice growing and provide about 70% of the rice. A kilogram of rice typically requires from 1500 to as much as 10,000 litres of water.

Japonica – One of the two main forms of *Oryza sativa*, the slightly sticky, medium grain japonica is the common rice of China, where it originated, and is also found in the cooler zones of the subtropics and in temperate regions. The grains are short, roundish, low shattering and have an amylose content which ranges from 0 to 20%. See also indica.

Jasmine – The fragrant long grain rice of Thailand.

Javanica – A third form of *O sativa* rice although not as common as indica or japonica. Grown on the Indonesian island of Java it's long grain like indica but slightly sticky like japonica to which it is more closely related.

Lemma – Outer layer of a rice husk or covering, see palea.

Lodging – The tendency for a rice plant to fall over due to its load of grain being too heavy for the stalk. Breeding a squatter, stronger 'semidwarf' rice was one of the keys to the development of IRRI's more productive 'new varieties'.

Long Grain Rice – Rice grains which, when cooked, are at least three times as long as they are wide.

Maturity – Developments in new varieties of rice have shortened the growing period to maturity from more than 200 days (for a very slow maturing rice) to as little as 90 days, enabling farmers to harvest more crops each year.

Medium Grain Rice – Rice grains which, when cooked, are 2-1/2 to three times as long as they are wide.

Milled – Once the husk has been removed rice is milled to remove the bran coating and turn the 'brown rice' into 'white rice.'

Moisture Content – Although the ideal moisture content varies from variety to variety if rice is too dry when it is ready to be milled it is likely to crack and broken grains are much less valuable than whole ones. On the other hand if the rice is too moist it will not store well and may go mouldy.

Node – As the rice plant grows the stem consists of a series of nodes and internodes.

Paddy – Rice that has been threshed but is still in the husk. The term comes from the Indonesia word padi and also refers to the field in which rice is grown.

Palea – Inner layer of a rice husk or covering, see lemma.

Panicle – The final branch of a rice stalk with the rice grains. A single stalk can have a number of panicles.

Patna Rice – Indian basmati rice from the state of West Bengal.

Parboiled – Unhusked rice which has been soaked and then briefly steamed before being dried, husked and milled. The rice retains slightly more protein and other nutrients than non-parboiled rice and the grains are also less brittle and less likely to break during milling. Parboiled rice is popular in Bangladesh, in southern India, Sri Lanka, the Middle East and in parts of Africa. In the US the branded Uncle Ben's 'converted rice' is parboiled.

Photoperiod – Some rice varieties are sensitive to the length of the day and do not flower until the sun is up for a certain number of hours, since that allows enough time for the rice grains to mature before the end of the summer. This type of rice would not be suitable for multiple harvesting so researchers try to develop rice varieties which are 'photoperiod insensitive.'

Pilaf – A flavoured rice dish popular in Turkey.

Polished Rice – See White Rice. The term is also used for rice which has been treated with talc, glycerine or a mineral oil to give it a glossy sheen; now an infrequent practise.

Puddling – Soaking and breaking down the soil so to limit the water loss through percolation. It was the development of the puddling technique along with transplanting which led to the major development of rice farming in prehistoric China.

Pulao – Indian or Central Asian version of the Turkish pilaf flavoured rice dish.

Radicle – first shoot that emerges from a germinating rice seed.

Rainfed – Non-irrigated rice grown in monsoonal regions where heavy and consistent rainfall usually guarantees that the rice will be kept wet. Nevertheless the often less than ideal water depth combined with overcast weather means that yields from rainfed rice fields are generally lower than for irrigated rice fields where the water depth can be kept at an ideal level and longer hours of sunshine promote faster growth.

Ratooning – growth of new tillers from nodes after harvesting, an alternative to burning off the old field, ploughing, flooding and seeding with new rice seedlings.

Red Rice – Rice with a slightly reddish or brownish colour, not to be confused with brown rice where the colour comes from the bran coating which all rice has until it is milled. Red rice is a speciality of the Camargue district of France but is also found in various Asian locations, particularly the Himalaya.

Rice Flour – Ground rice grains used as a food additive in products such as baby food or snacks.

Rice Paper – Thin sheets of rice and water paste which is first dried then rehydrated to make a wrapper for Vietnamese and Thai spring rolls. The 'rice paper' used for writing or drawing on, rather than eating, is a misnomer since it is not made of rice at all.

Rough Rice – See Paddy Rice.

Sawah – An individual rice field in Bali and other parts of Indonesia.

Semidwarf Rice – The development of high yielding rice varieties in the 'Green Revolution' which dramatically increased world rice production in the 1960s, depended on developing a short, sturdy rice with so many grains per stalk that it would have fallen over if it had been taller. Modern high yield rice varities are mainly 'dwarf' or 'semidwarf' rice averaging around a metre high versus 1.5 metres or more for older rice types.

Short Grain Rice – Correctly rice with grains less than twice as long as they are wide but the term is also used to refer to anything shorter than long grain rice.

Slash & Burn – Primitive nomadic form of upland rice growing where vegetation is slashed down and burnt before the rice is planted. The soil fertility is quickly exhausted.

Starch – About 80% of a grain of rice is the white carbohydrate known as starch or technically as amylaoe and amlyopectin.

Sticky Rice – In Japan the favoured rice is a short grain variety which tends to stick together when cooked. It's highly suitable for eating with chopsticks or for making sushi.

Subak – The rice growers co-operative which every rice farmer is a member of in Bali, Indonesia. The subak's most important role is to ensure equable distribution of water resources in the traditional irrigated fields in Bali's hill country. Since water has to flow downhill from field to field it's said the best person to run a subak's water distribution programme is the farmer at the bottom of the hill.

Swidden – See Slash & Burn.

Tiller – The uppermost node of a rice plant stem bears a leaf and a bud which can produce a tiller. Tillers growing out from the main stem are primary tillers which may in turn produce secondary and tertiary tillers. Rice breeders have traditionally tried to grown rice plants with numerous tillers – a rice plant with good 'tillering' ability. Recently attention has turned to producing plants with fewer tillers but more grain per pannicle or with fewer tillers so that more rice plants can be grown per square metre.

Transplanting – In Asia rice is traditionally transplanted rather than 'broadcast'. A small nursery of seedlings is grown in the corner of a field and when these seedlings are a few cm high they are transplanted into neat rows in the main field. The advantage of transplanting over broadcasting is that the rice is already of a reasonable size before any encroaching weeds can grow around it and the neat rows permit easy access for weeding and fertilizing. The disadvantage is that transplanting is labour intensive and very hard work. Modern machines have been developed and are used in places like Japan with high incomes but small fields suitable for transplanting.

Upland – Hill country rainfed rice fields where yields are usually quite low. Slash & burn agricultural techniques are often followed in these areas.

Water – Rice needs lots of water, typically growing one kilogram of rice requires 3000 to 10,000 litres.

Wild Rice – Although closely related, *Zizania aquatica* or *Zizania palustris* is not actually rice at all, it grows wild naturally only in the shallows of some lakes in North America. This 'wild rice' is actually cultivated but the term is also used to refer to wild varieties of cultivated Asian rice.

White Rice – Rice grains which have been husked and milled, removing the bran coating from the grains.

Whole Rice – See Brown Rice.

Yield – Amount of rice grown on a given area of land in one harvest season. Typically measured in tonnes per hectare, yields can be under one tonne per hectare for deepwater rice or upland rice, through three to five tonnes per hectare in many third world environments to more than 10 tonnes per hectare in Australia and the USA where maximum advantage is taken of modern farming techniques.

Rice-Planting Song

Plant-ing rice is ne-ver fun, Bent from morn till the set of
When the ear-ly sun-beams break, You will wond-er as you a-

sun, Can-not stand nor can-not sit Can-not
wake, In what mud-dy neigh-bor - hood There is

rest for a lit-tle bit. Plant-ing rice is no
work and the plea-sant food.

fun. Bent from morn till the set of sun; Can-not

stand, can-not sit, Can-not rest a lit-tle bit. Oh,

come friend, and let us home-ward take our way,

Now we rest un - til the dawn is gray,

Sleep, wel - come sleep, we need to keep us strong.

Morn brings an - oth - er work day long.

Philippines – The rice-planting song – Riceworld Museum at IRRI in Los Baños.

THANKS FROM US

From basic research to rice on the table we were given a huge amount of help every step along the rice trail. In particular we have to thank the staff of IRRI, the International Rice Research Institute, not only at their headquarters at Los Baños in the Philippines but also at IRRI regional offices and other associated rice research stations. Moving alphabetically around the region our heartfelt thanks to:

Australia – Kerry and Nick Lowing who introduced us to high tech rice farming at Moulamein in New South Wales. Ruth Arthur and her husband, the late Neil Arthur, opened their home to us at Moulamein in New South Wales. Jacqui Herrman, working at the time with the Ricegrowers Co-Operative, organised Richard's visit to the Rice Festival in Leeton and Griffith. Subsequently Claudine Menegazzo at Ricegrowers has been very helpful. Laurie Bisa flew Richard over the rice bays in a hot-air balloon.

Bangladesh – Dr Sadiq I Bhuiyan, Salim Ahmed and Jamila Khandekar from IRRI in Bangladesh and Atiqul Islam and Abdul Muttaleb from the Bangladesh Rice Research Institute introduced us to the fascinating story of deepwater rice in Bangladesh at Daud Kandi and Tangail.

Burma – U Hla Min was our guide and interpreter when we chased rickshaws around Rangoon for our earlier book on those fascinating bicycle-powered taxis. This time he introduced us to the rice business in Rangoon, the capital, and travelled with us out to Pathein in the country's Irrawaddy Delta ricebowl. We also enjoyed his company on a sidetrip to the balancing pagoda of Kyaiktiyo. In Burma we were also welcomed to the Rice & Paddy Wholesale Depot in Rangoon port, sat in on the rice trading activities of OK Trading and watched the goings on at U Myo Wai's wonderful old rice mill at Pathein. We were also very impressed by the amount of rice their customers consumed at Nila Briyani in central Rangoon.

Cambodia – Dr Peter Cox, Dr Harry Nesbitt and Dr Gary Jahn, with IRRI in Phnom Penh, told us the remarkable story of the rebuilding of rice farming in Cambodia after the devastation of the Pol Pot years. Dr Nesbitt has been dubbed the 'Rice God of Cambodia' for his amazing success in restoring the rice growing abilities of the war torn country.

China – In Yangshuo Wang Ren Yong, better known as Uncle Bob to Yangshuo backpackers, teamed us up with Mo Jiang Ming, the best guide anybody could ask for. Together we explored ricelands and rice markets around Yangshuo and then made an excursion to the dramatic Dragon's Backbone Rice Terraces. Mr Fan was our calm and collected driver. We met some remarkably friendly and welcoming rice farmers as we travelled China's rice trail.

India – Our IRRI contacts in India led to a full faculty of rice scientists at first the Kaul Rice Research Station in Haryana, then the Saline Soil Research Centre in Karnal and several other Haryana and Punjab research centres, and then to the Ghaghraghat Research Station in Uttar Pradesh. Our grateful thanks to Dr S D Dhiman, Dr D S Dodman, Dr Rakesh Sharma, Dr Khushi Ram, Dr B Mishra, Dr R K Singh, Dr T S Bharaj, Dr Malik Singh and Dr Ajmer Singh. We were delighted to discover when we checked in to our hotel in Karnal in Haryana that we had also become Doctors. Our driver in Haryana was the ever helpful Rachhpal Singh. On the other hand our driver in Uttar Pradesh provided us with two heart stopping tyre blowouts, after the second one we abandoned him by the roadside and took a bus back into Lucknow.

Indonesia – Gede Lata, losmen (a small Indonesian inn) manager at Tirtagganga, was our entrée to the harvest at Ababi in Bali. Nyoman Sutawasa was both driver and guide as we travelled around the Island of the Gods. Ketutu Suartana, a long time friend in Ubud, also helped us track down interesting rice tales.

Japan – The tatami mats were rolled out for us in Japan thanks to IRRI's Dr Hiroyuki Hibino and his assistant Mrs K Morooka. Mrs Mitsuko Hibino drove us around the Tsukuba area and introduced us to local rice farmers and then accompanied us by train to Morioka to see more traditional rice farming and a rice festival. Mrs Hibino's tales of catching crickets in the rice fields when she was a child were followed by a bowl of fried crickets with our ryokan meal in Omagari. Thanks also to her friends Hiromi Yoshino and Toshi Sawaguchi. The Hibino's son Taisuke Hibino was also very helpful. Spending time at Yoshio Otani's traditional tatami mat manufacturing business was particularly interesting.

Nepal – Much of our time in Nepal was spent gazing morosely at rainclouds when the monsoon should have ended weeks earlier. The clear blue skies and Himalayan skyline we needed finally arrived only hours before our flight departed and we checked in still kicking the paddy field mud off our shoes. Our patient guide for our wet and muddy Kathmandu Valley excursions was Adwait Pradhan.

Philippines – At IRRI's Los Baños headquarters we stayed at the IRRI guest house and were introduced to the institute's first rate work by Ian Wallace and Gene Hettel. Aurora Hettel travelled with us up to Banaue to see the famous rice terraces and to Lucban to witness the annual Pahiyas Festival.

Thailand – IRRI's local office again came through for us with sterling help from Dr Boriboon Somrith, Sutep Limthongkul, Weerasak Sri-oun and Jutharat Prayongsap. Our visits to the rice research centre at Pathum Thani (thank you Orapin Watanesk) and the Prachinburi centre, with its deepwater rice expertise, were of particular interest. We especially enjoyed the time spent with the happy harvesters we encountered near Pathum Thani but we were also impressed by the rice processing prowess of the staff at Riceland International in Ayuthaya and the rice bidding wars we witnessed at the Suphanburi Agricultural Central Market were some of the most exciting moments along the rice trail. Our rice mill visits were facilitated by Sawang Sriprasert, Nantawan Jaided and Pairoj Jumpar-Ngern.

Vietnam – In Vietnam we travelled by ferry to Cantho in the Mekong Delta where Ng Ng Ngoc Huy of Canthotourist shuttled us around the rice markets, rice farms and rice processors of the region with efficiency and friendliness. His friend Maurice Cao Duong My drove us back to Saigon making more rice stops along the way.

Everywhere – In every country along the rice trails rice farmers, rice processors, rice traders and rice consumers made us welcome and told us about their part in the story of rice. Gene Hettel at IRRI read through the nearly completed text and pointed out the worst errors, if others have subsequently crept in that's our fault, not their's.